TARGETED:

"If I Die, This Program Killed Me!"

Renee Pittman

Copyright © 2024

Mother's Love Publishing and Enterprises

All rights reserved.

ISBN: 978-1-7374060-0-6

DEDICATION

To the survivors, who refuse to bend,
or break, speaking up and blessing others

TABLE OF CONTENT

Acknowledgment 6

CHAPTER 1 Workers of Iniquity 1

CHAPTER 2 "Shadow of The Almighty" (Psalm 91:1).........23

CHAPTER 3 The Denizens of the Abyss Call Their Names.38

CHAPTER 4 Falling into the Abyss Has Never Been the Fate of Those With Wings 61

CHAPTER 5 LAPD "Rampart Scandal" Act II.......................77

CHAPTER 6 Organized Stalking is No Delusion 104

CHAPTER 7 The Blog 'THEY' Don't Want You to See!......118

CHAPTER 8 Professional Gaslighting...................................248

CHAPTER 9 The Mechanics of Detecting Voice to Skull (V2K) Official Beamed Schizophrenia 353

CHAPTER 10 If a Wheel Squeaks Loud Enough it Will Get Oiled 373

About the Author 407

Acknowledgment

To many nationwide and globally enduring the
unimaginable, while fighting the good fight, with courage
and heartfelt determination that
originate from brave hearts.

Thank you!

CHAPTER 1

Workers of Iniquity

The dog and wolf represent that which comes from the deep. It is the nameless, hideous tendencies in humans that are in sync with that which dwells below.

Between the two dogs runs a pathway with a tower on each side. This pathway symbolizes the human journey from consciousness to the depths of unconsciousness. It represents the becoming of animal essence that strives for manifestation, becoming greedy for it, which always involves a pack, a band, a population, a peopling, or a multiplicity.

The lobster crawls from the abyss of water to land. It is a creature that lives on the bottom of the ocean yet emerges out of it and is associated with the liminality of consciousness and unconsciousness that ultimately sinks back whence it came based on bad choices.

The checkered black and white patterns allegorically represent polarity, good or evil, right or wrong. It shows us that life is not ideally defined and that we need to define for ourselves the right path to choose. There is a mystical, esoteric meaning that can apply to all areas of life, with good and evil being a very common dichotomy within the hope that life's experiences will bring forth consciousness evolution.

And, it is vital to remember that if you anticipate the command; you are out of step.

***

I am tired of your mouth rang out inside my home through the beamed communication system directed at me from one of several official setup locations, with drones positioned overhead a few days before Thanksgiving Day 2022. When I heard the beamed Audio Spotlight harassment, I laughed at the ludicrous threat. I laughed at the foolishness of men sitting behind advanced technology and their unified obvious mentality of deceit who had shown up around me now entering 18 years and toyed with me heinously, thirsting to devour me. The game plan was a hope to destroy my self-esteem first, hoping to do so, by degradation and beamed torture which many officially targeted with patented technology, targeted from sunup to sun-down, report experiencing nationwide. The comment from the DOD contractor, operating from this specific location, demonstrated a disturbed mind. I thought "I'm tired of you ALL too!"

As drones hovered above brilliantly using patented mind-invasive technology, reading subvocal thoughts and looking for fear and weakness, along with military-beamed assaults, I thought, "Thank you for the compliment". Was I supposed to make their high-tech targeting easy for them? I think not and attempt to prove this every day.

The operators of this technology today are a hideous crew of official personnel who have, since I moved in 10 years ago, gradually taken permanent residence in several compliant locations around me, becoming a permanent fixture. The bonus for homeowners is free mortgage payments, courtesy of Uncle Sam's nationwide billion-dollar "Black Budget" secretively financing ongoing human experimentation programs across the USA focused on individuals, groups, communities, and large populations. They were testing the progression of many advanced technologies with mind hacking and hive-mind abilities.

For 18 years, as of 2024, I had been on the receiving end of a massive effort to permanently silence me, backed by lies told to motivate and mobilize the community as hired stalkers. The constant bullying and fear monger death threats were not only to me but threats to harm my family. The silencing effort, right out of the Psyop playbook, also included many typical COINTELPRO entrapment schemes. Naturally, if something like this is happening across the USA it has to be shrouded in secrecy.

Official government agencies are well aware of and spearheading the high-tech military unification with civilian agencies and technology such as the military Active Denial

System "Pain Ray" used for subjugation, and also the use of various electromagnetic weapons overall.

Night after night, the hardware in my right ankle, fixed with 13 pins and screws and two titanium plates on both sides, was hit, the result of a 2007 car accident, directly related to this targeting program. Cooking the ankle and other areas of the body using the "Pain Ray" left me limping in pain inside my house. The torture intensifies, with my obstinate resistance and goal of hoping a target will snap.

When connected, Military COINTELPRO Intelligence operations unite with unpublicized, top-secret sections of nationwide Fusion Centers, and the realization of the imminent threat of exposure, accurately, of official criminal activity among the agencies involved, they come out in droves. Many remember that MK ULTRA mind control studies included various agencies' unification, both official intelligence, military, and civilian.

Patented mind-reading technology, reading brainwave patterns, is the key to human experimentation, which I document in other books in this series. Without this capability, this program would be fruitless.

With military personnel at the helm of behavioral modification and health-deteriorating weapons, along with thought deciphering, the awareness that targets are figuring out they are being ruthlessly used sounds an alarm to those with a vested interest in ongoing human experimentation, who are destroying lives ruthlessly, triggering them to embark on various diabolical silencing schemes.

With awareness, targets are ushered into a new phase. In this phase, the beamed torture and verbal harassment begin and quickly escalate. This is to counter the truth when targets speak up. The goal is to ensure people are automatically deemed mentally ill as the strategic machination unfolds using any and everyone, including gullible Psyops, which is a modernized version of Stasi Zersetzung tactics. With some, the ones able to become highly effective exposing this program, this can be the final stage for a strategic, high-tech death sentence that then encompasses various efforts designed to first get targets to sabotage themselves. This increases the perception of mental illnesses of anyone speaking up. With each failure to stop exposure, and ego-driven anger of those involved, and after many entrapment schemes are ineffective, this program will set the stage for cleverly crafted death.

These are highly effective tactics played out and perfected over decades of human experimentation. When thousands, harmless, use the term "Targeted Individuals" step forth saying they are being used, as human lab rats, drastic measures are taken, within the coverup, beginning with various labels, paranoid, schizophrenic, delusional, and "Mass Delusions".

Official manipulation and influence tactics develop, using patented technology designed for behavior modification, which can, with some horrific, result in sacrifices and casualties when pushed over the edge completely broken and hurt self and others. Human lab rats become angry, and out of control about the destruction of all aspects of their lives, and relationships, including family, friends, and love interests.

It must be understood that it is a small few versus thousands who have not broken. The small few allow strategic publication and news articles that claim that ALL targets, and those labeled Targeted Individuals, are mentally ill and, more importantly, a danger to self and others. In this diabolical orchestration, again, a small few are nudged to create atrocities by unyielding harassment. When they do, they live out the expression, "Crazy is as crazy does" and it becomes fact, and there are no excuses. Only the strong survive the high-tech targeting and many have survived upwards of 30 years.

If official personnel can make the target snap before targets open their mouths reporting being used for experimentation, they can discredit everyone and this allows this program to continue, unscathed using various types of brilliant mind-invasive and psychophysical electromagnetic weapon technology on them and anyone else. Historically, this program has no plans to shut down ongoing research documented as generation to generation.

Again, the first order of business is the mental illness tag documented in hospital records resulting from telling anyone, and specifically hospital staff, when detailing what the target is experiencing. The problem today is that these operations increasingly cannot control the explosion of the truth all over the world as people wake up.

For example, let's use Assistant District Attorney Myron May, Florida State University, Strozier Library shooter, November 2014, as an example. His targeting sped up fast, due also to his connections, intellect, and credibility before it went heinously south for him.

Of course, highly intelligent individuals figure out they are being used and experimented on quickly, and as awareness looms, understand that they likely have been part of this specific type of experimentation, nonconsensual, since childhood. This is especially true with gifted children of all races and recognition at any time.

The result is that agencies with a vested interest in keeping the truth hidden to avoid backlash and disgrace show up in full force, not only convincing everyone around the target that they are crazy, but gaslighting the actual target of such. This sends the target to the hospital and thus the medical record misdiagnosis is set. EVERYONE's biometric signatures have been uploaded into the supercomputer database resulting in technology where anyone can be tracked, monitored, and influenced at any place on the face of the Earth due to the brilliant military space-based system. It takes just three satellites to blanket the Earth with detection ability and thousands orbit the Earth.

Another scenario that is popular by operatives, allowing this program to come into the open, are reports that targets did something that deserves the targeting, and the vicious destruction of their lives. This is what is also used to mobilize the public as an arm of harassment.

The setup is so diabolical that fictitious reports are promulgated to communities, Citizen Volunteers, and InfraGard. An investigation is now in the hands of law enforcement, using military psychophysical technology with military personnel, making human experimentation legal, whose goal is to nab the target and who couldn't care less about how and why the person is targeted. As this program, now unified or "fused" with civilian agencies, federal agents, and military intel overseen, cops are provided advanced beam assault, mind invasive, and behavioral modification technologies which they use to try to create the goal they seek including the use of patented subliminal influence tech. The program today results in all levels, federal, state, and local authorities, now approved for military-grade technologies, working with military personnel for mind control and psychophysical experimentation which in reality is designed to keep the truth hidden of advanced technology awareness from publication within the public domain.

Documented in "Covert Technological Murder…" are two surgeries for me as a result of beamed assaults and counting. The focus is now on my knees for knee replacements. I have documented the strategic slow deterioration of my knees in my medical records as coming with my left knee first. When this program is set up in your neighborhood you are marked for clever high-tech demise, with beamed pain and suffering that increases. You are a challenge when you cannot be broken and on top of this, you have the gall to expose them as official sociopaths running a horrid program.

As far back as Nikola Tesla in the U.S., and his early 1900s Death Ray, scientists have investigated the electromagnetic energy force that can be used as a weapon of war. Because of this, it is little wonder that the idea of directed-energy weapons soon began to capture the attention of military planners. Much of the defense-related research has been focused on how to fashion electromagnetic energy into powerful, precise beams capable of creating militarily useful effects.

This research explores the current capabilities of the US military to use electromagnetic (EMF) devices to harass, intimidate, and kill individuals and the continuing possibilities for violations of human rights through the testing and deployment of these weapons.

The pain ray is one of the most painful beams a person can experience, which the Department of Defense believes is the perfect weapon for controlling others. It leaves no evidence, in darker skin, however, red burn patches in people lighter. If a target's legs are the focus, when the target moves and muscles flex, it is unbearable pain.

Weapons classified under the general heading of directed-energy weapons (DEW) include high-energy lasers, electromagnetic railguns, and radio frequency weapons (high-power microwaves or ultra-wideband weapons). Over the past decade, the technical maturity of these weapons has accelerated, and what was once a "promise" is now emerging as reality. The Office of Naval Research (ONR) is developing and working to scale up the power of free-electron lasers, chemical lasers, and their associated beam directors, radio-frequency weapons, and full-scale electromagnetic rail guns capable of launching precision-guided hypersonic projectiles.

Typically, after a short time, the dielectric heating pain to the skin will subside. However, it is being used relentlessly today on Targets. Although it is listed as nonlethal it can not only cause great pain as it sears into the skin painfully heating the top layer. It also causes psychological pain with the realization that another person is sitting at the helm and sadistically doing this to you. It is highly effective because people do not want to suffer and some will give in hoping it will stop doing whatever it takes. Typically, with official personnel at the helm, the request is some type of demand or action.

Knees become the perfect target area for deterioration because they are covered in a thin layer of skin where beamed assaults can do maximum damage quickly. This program knows this, likely by crippling many across our nation watching in real-time the outcome. When you go into poor communities, many are limping, in wheelchairs, etc. Poverty-stricken communities have always been historically and today the first stop for various vicious types of human experimentation.

In my case, three of the locations set up around my neighborhood are federal agents, military personnel and DOD contracted trainers training everyone, including corrupt Black cops dispatched from the Los Angeles Police Department.

Official recruitment is vital within the community as neighbors' witnesses, because of the use of their homes allowing the setups, and top-secret use of bio and psychotronic systems and devices.

The fact is, what is happening today, was intentionally legalized specifically for military and law enforcement, Post 9/11, for use of electromagnetic weapons. Legalization was approved for crowd control, for police investigations, which are bogus and proven by the subjugation torture, and for military research activity.

In September 2006, Air Force Secretary Michael Wynne announced that crowd control weapons should be tested on Americans first. "If we're not willing to use it here against our fellow citizens, then we should not be willing to use it in a wartime situation," said Wynne.

Psychotronics is the ability, using extremely low frequency (ELF) waves, and other advances to manipulate a person's emotions, thoughts, bodily functions, and all from a remote location. Using this technology, a person or an entire population can be effortlessly controlled like robots with remote control also known as Remote Neural Monitoring, and, as the technology continues to advance connected to a

Brain-Computer Interface (BCI) or Artificial Intelligence (AI) and again, Hive Mind experimentation.

The Earth is wrapped in a donut-shaped magnetic field. Circular lines of flux continuously descend into the North Pole and emerge from the South Pole. The ionosphere, an electromagnetic-wave conductor, 100 kms above the earth, comprises a layer of electrically charged particles acting as a shield from solar winds. Natural waves are related to the electrical activity in the atmosphere and are thought to be caused by multiple lightning storms.

Collectively, these waves are called "The Schumann Resonance," the current strongest at 7.8 Hz. These are quasi-standing ELF waves that naturally exist in the earth's "electromagnetic" cavity, the space between the ground and the ionosphere. These "earth brainwaves" are identical to the spectrum of our brainwaves. (1 hertz = 1 cycle per second, 1 khz = 1000, 1 mhz =1 million. A 1 Hertz wave is 186,000 miles long; 10 Hz is 18,600 miles. Radio waves move at the speed of light.)

As detailed in another book in this series, the Creator designed living beings to resonate with this natural frequency pulsation to evolve harmoniously, which is being used for people to evolve inharmoniously. As science progresses, these natural geomagnetic waves are being replaced by artificially created very low frequency (VLF) ground waves initially coming from, for example, old GWEN Towers.

As detailed in Book I, "Remote Brain Targeting," Dr. Andrija Puharich (in the 1950 & 60s), found that a clairvoyant's brainwaves turned to 8 Hz when their psychic powers were operative. In 1956, he observed an Indian Yogi controlling his brainwaves, deliberately shifting his consciousness from one level to another. Puharich trained people via bio-feedback to do this consciously, that is, creating 8 Hz waves with the technique of bio-feedback. A psychic healer generated 8 Hz waves through a hands-on healing process, actually easing that patient's heart trouble; the healer's brain emitting 8 Hz.

One person, emitting a certain frequency, can make another also resonate to the same frequency. Our brains are extremely vulnerable to any technology that sends out electronic frequency waves because they immediately start resonating to the outside signal by a kind of tuning-fork effect.

Puharich further experimented, discovering that 7.83 Hz (earth's pulse rate) made a person "feel good," producing an altered state * 10.80 Hz causes riotous behavior. 6.6 Hz causes depression, 10 Hz puts people into a hypnotic state

Puharich and American Dr. Robert Becker (The Body Electric) designed equipment to measure these waves and their effect on the human brain. Puharich started his work by putting dogs to sleep.

By studying and modeling the human brain and nervous system, the ability to mentally influence or confuse the human mind is also possible. It is possible to create depression and a feeling of overwhelming hopelessness by the induction of a current into the electrical circuit of the brain. Through sensory deception, it may be possible to create synthetic images or holograms to confuse an individual's visual sense or, similarly, confuse the senses of sound, taste, touch, or smell.

Our bodies operate much like a computer. It contains a vast number of neurons that can be likened to a data processors. As an example, the computerized human biological system uses chemical-electrical activity of the brain, heart, and peripheral nervous system. During our daily life, our bodies process the signals sent from the cortex region of the brain to other parts of our body. Like a well-oiled machine, which also impacts the tiny hair cells in the inner ear that process auditory signals, and also the light-sensitive retina and cornea of the eye that process visual activity.

Because of this the human body, bioelectric is capable not only of being deceived, manipulated, or misinformed, it can also shut down or destroyed operating just like any other data-processing system.

Using the electromagnetic frequencies, the "data" the body receives from external sources, for example, microwaves vortex, or acoustic energy waves, created through its own electrical or chemical stimuli, the human body and mind can be manipulated or changed just as the data (information) can be in any hardware system can be altered.

With the patented capability to monitor subvocal thought, (mind reading) when we say words to ourselves, silently, or read a book, we can feel the slight sensations of those words in our vocal muscles - all that is absent is the passage of air. Mind or thought reading is an enhanced version of computer speech recognition, with EEG brainwaves being substituted for sound waves. The easiest "thought" reading is remote picking up of the electromagnetic activity of the speech-control muscles. Advanced research by scientists has also found that they could convert thoughts into computer commands by deciphering the brain's electrical activity.

Naturally, technology can capture the thoughts and minds of the global population, and control both interests of those seeking to do so. President Lyndon Johnson's Science Adviser, Dr. Gordon J.F. MacDonald wrote the globalist-promoting 1968 book, unless peace comes, a scientific forecast of new weapons. Within MacDonald described how man-made changes in the electrical earth-ionosphere can be used for mass behavioral control. He said that low-frequency electromagnetic oscillations can attack the low-frequency electromagnetic brain waves in human beings.

Zbigniew Brzezinski, the founding Director of David Rockefeller's Trilateral Commission, also served as President Jimmy Carter's National Security Director, where he founded the infamous FEMA that is reportedly designed, as a conspiracy, to implement world government dictatorial rules over the U.S.A. predicted these exact types of electromagnetic psychotronic weapons.

He stated: "It is possible - and tempting - to exploit, for strategic-political purposes, the fruits of research on the brain and human behavior. Accurately timed, artificially excited electronic strokes could lead to a pattern of oscillations that produce relatively high-power levels over certain regions of the earth. In this way, one could develop a system that would seriously impair the brain performance of a very large population in selected regions, over an extended period."

The fact is Extremely Low-frequency radiation affects the electrical activity of the brain and can cause flu-like symptoms and nausea. Other projects sought to induce or prevent sleep, or to affect the signal from the motor cortex portion of the brain, overriding voluntary muscle movements. This is the foundation for Brain-Computer Interface (BCI) technology advancing today.

There has also been extensive research into high-tech telepathy research and the beamed "Schizophrenic Effect" within the USA and U.S.S.R. The research has focused specifically on manipulation of emotional or behavioral impulses which today is in widespread use and something that the agencies involved do not want known or publicized. These weapons have advanced to a highly perfected state that influences telepathically the behavior of people, alters their emotions or health, and even kills at long distances, leaving a trail of deadly human experimentation for decades.

"Government authorities and the military would have to overcome no insurmountable difficulties to modulate carrier frequencies with ELF signals in existing centimeter radio-relay links...an army of occupation could then manipulate a nation's ability to make decisions." Subliminal messages bypass the conscious level and are effective almost immediately.

The ability to cause death can be transmitted over distances, thus inducing illness or death for no apparent reason. These weapons would be able to induce illness or death at little or no risk to the operator. The psychotronic weapon would be silent and difficult to detect which is the beauty of the criminally insane.

One of the most prolific DOD Contractors who has focused on advancing Directed-energy weapons for many years which are currently being deployed included a microwave weapon manufactured by Lockheed Martin and used for a process known as 'Voice Synthesis' which is remote beaming of audio (i.e., voiced or other audible signals) directly into the brain of any selected target.

Satellites, controlled by advanced computers, can send official personnel voices in every language, where the operator's perceptions and influence are effortlessly inserted into a person's subvocal thoughts. On a massive level these space-based systems can target any population of choice with this diffused artificial thought process system. Again, the chemistry and electricity of the human brain can be manipulated by satellite and even suicide can be subliminally influenced. Today the U.S.A. and Russia have the highest numbers of satellites orbiting Earth with China fast approaching.

One of the primary goals of mass mind control and intense focus on groups and populations is disruption and chaos. The ability to influence emotional anger, as well as ferocious, anti-humanitarian behavior, or extremist groups can be fabricated effortlessly. Without a doubt, the resulting activity in some cases deadly disruptions can be instigated by advanced satellites using bioweapon psychotronics electromagnetics in many countries in Asia, Africa, Europe Latin America and naturally America.

Scientist confirm that these systems and BCI computers have been fed with the world's languages, enabling synthetic telepathy to reach into people's heads making people believe God is speaking to them personally or for mass population control enactment of the Second Coming, complete with holograms! Today, we have intellectual, physical, and financial resources to master the power of the brain itself, and to develop devices to touch the mind and even control or erase consciousness.

Using two-way radio communication, called remote neural monitoring, a wavelength can be brain transmitted into a person's head. The wavelength streams through the brain and returns to a computer, where all aspects of a human being's life, including thoughts of every kind, are uncovered and analyzed. This is why I have often said that without the ability to read a person's mind, mind control experimentation would be fruitless. By electrical, electromagnetic stimulation of specific cerebral structures, radio command can induce movements. Emotionally disturbed personalities can be created that can appear or disappear, social hierarchy can be modified, sexual behavior or preferences can be changed, and memory, emotions, and the thinking process may be influenced by remote control and without electro-lode transmitters.

In the newspaper Delovoi Mir, "Mind Control" by Ivan Tsarev, 1992, a victim wrote,

"They controlled my reactions, my thoughts, and caused pain in various parts of my body...It all started in October 1985, after I had openly criticized the first secretary of the City Committee of the Communist Party.

The symptoms of psychotronic weapons common to most human guinea pigs across the USA and globally, reported 24 hours 7 days a week, for years on end, are all kinds of harassment using patented technologies. Scientifically, the technology can remotely target and control every nerve of the body. Heart rate can speed up and slow down, bowel movements can be regulated, and illnesses can turn on and off in an instant. Victims report microwave hearing or voices in the head Voice to Skull (V2K) and sleep deprivation. Thoughts can be read, and played back to the victim, instantaneously. People around the victim can repeat verbatim, the victim's immediate thoughts or the beamed thoughts of official personnel at the helm.

There appear to be standard procedures and officially approved methods fueling human experimentation:

1. behavior modification and the mind subliminally influenced and controlled,

2. homosexual and degrading themes vastly reported to also influence actions and change sexual identity,

3. emotions are played on ridiculed and denigrated

4. high-tech pain creating trauma-based mind control where the pain starts and stops briefly and the targeted thinks it is over and free, then the beamed pain intensifies and starts repeatedly,

5. dream manipulation,

6. remote sexual manipulation, stimulation and abuse with pedophilia ideation, rape setups while everything is real-time monitored,

7. drug bioweapon testing addiction influence,

8. destructive thoughts to influence hurting self or others,

9. microwave and tissue burn,

10. bizarre stalking tactics used for testing psychological operations,

11. electrical technology manipulation, phone, car, TV and computer,

12. official terrorism such as slashing tires, break-ins, email and mail tampering,

13. death threats to incite fear

14. isolation,

15. visual hologram projection.

According to victims, the experimentation is horrific, ruthless, vindictive, vicious, amoral, inhumane, sadistic, and cruel. Most victims describe the experience as very debilitating and liken it to mental rape, and an official prison designed for total destruction of the quality of life. Again, everyone is labeled mentally ill and lives with financial ruin, and loss of health, social life, and career. All say, and we are talking thousands speaking up today, that the technology is so advanced, sophisticated, highly effective, and portable that it results in the expert, high-tech deterioration of mental, emotional, and physical health and gradually slow death.

Unfortunately for the victims who are being used as guinea pigs for this technology as it advances (who will someday most likely be you), many are left helpless as law authorities and government agencies refuse to investigate or cooperate towards ultimate social and mass population control secret.

The Sound of Silence is a military-intelligence code word for certain psychotronic weapons of mass mind-control tested in the mid 1950's, perfected during the 70's, and used extensively by the "modern" U.S. military in the early 90's, despite the opposition and warnings issued by men such as Dwight David Eisenhower. This early developed mind-altering covert weapon is based on subliminal carrier technology, or the Silent Sound Spread Spectrum (SSSS), (also nicknamed S-Quad or "Squad" in military jargon.) which I have documented again as prominent in the other books in this series. It was developed for military use by Dr. Oliver Lowery of Norcross, Georgia, and is described in US Patent #5,159,703.

Wirtschaftswachstum verspielt: **On Perceptual Economy: subversive tendencies in simulations is defined below...**

From: Vladimir Muzhesky

From Psychotronic Warfare to Biotronic Materials (excerpt)
Below is official documentation:

Vladimir Muzhesky

Synthetic Plane of Immanence

Whether electronically or materially mediated, psychotronic complexes interact with their spatial background environment and not with their generating concept. The concept itself is rerendered into a spatially invaginated structure intended to reflect and replicate elements of abstract defense. It is a multipolar semiotic system that goes beyond the tripolar systems of Lotman, more reminiscent of strange attractors: it defines multi-dimensional landscapes of bio-informational trajectories instead of describing cultural codes. In general, instead of being encrypted in bio-informational activity, culture, with its codes being hacked vertically and horizontally through mental manipulation which disperses personal and collective memory, replicates and counterfeits reflections and intentions and even meaningful positions, then mutates into a new dimension where meaningful and nervous spatialities are fused to form a twisted conglomerate of neuro-space: a virtual plane of immanence for the economy of abstract defense.

It would not be fair to say that cyberspace was out there first. It always was and forever will be a model which claimed new dimensions. Once we speak about modes of existence in this dimension we cannot avoid a correlation with biological informational networks: hence a cyberspatial dummy will never be actualized as in the neomythology of sci-fi literature, but neutralized, as happened with the defense technology in the ex-Soviet Union.

***

The global Cabal's agenda is part of the New World Order agenda of evil to keep experimenting and fine tuning these technologies separately to control and to spy on Americans and people all over the world while mentally controlling everyone, simultaneously. Without these devices, and global space-based systems, that connect on a basic level such as smart phones and Wi-Fi providing desirable features of convenience, few would use them, and the Cabal would not be able to fine tune these technologies and all connected to the Earth's Electromagnetic Spectrum.

Once the fine tuning is complete, reportedly the Cabal has planned to combine all forms of systems, under control of a massive Quantum computer using A.I. automatic controls and operating through inter-dimensional wave forms. At this point, they believe they will be able to make whole populations completely mind-controlled and cloned/hived mind to the mother A.I. computer; where they can make people sick or even murder whole hum

Reportedly, they plan to eliminate 90% of all humans and replace them with their own synthetic race of hybrids, trans human of a combination of clay and iron (biology and manufactured electrical components). Whether fact or fiction, one thing is certain, the human soul is infinite but theirs is not which is their motivation to survive and never die. Their time is running out!

CHAPTER 2

"Shadow of The Almighty" (Psalm 91:1)

In one of the most uplifting passages of Scripture, we have this reassurance: "He who dwells in the shelter of the Most-High will abide in the shadow of the Almighty" (Psalm 91:1, ESV). To dwell in the shadow of the Almighty is to live under the promise of God's protection.

I will never forget being called to the principal's office and told that someone wanted to speak with me and run tests. There stood a young white man in his early 30s. I followed him into a small room next to the Principal's office. The tests were designed to test intellectual ability and ability to figure out various mental challenges and with me, it was lightning fast.

Some states call this type of testing, and results, Talented and Gifted, however to this day in Los Angeles, it is still called Special Enrichment. I remember what initiated the testing. Two weeks prior during math study time, in a class of 25 students, we were given a challenge to see who could complete approximately 30 multiplication problems first. The teacher explained the details and then set the timer for 30 seconds. I looked across the room and there sat my elementary school crush, Andre Greer, knowing he would be my competition.

The buzzer began the countdown and within minutes I had whizzed through each problem, then announced, 'finished' with 100% percent accuracy seconds before the buzzer stopped. Andre Greer followed. After this, a person, a psychologist, arrived. Later in life, I learned America was involved in studies to determine if White people were/are more intelligent than Black people or were Black people simply physically stronger and predisposed to lower intelligence. Now sitting with him and completing the challenges he presented, I thought he would be pleased that I had similarly completed his test as I had the math problems. However, when I looked into his eyes, there was no delight or happiness. I will never forget the look that shattered his preconceived notion of inferiority, testing all schools and children of all races. The fact, based on his negative reaction, I had likely outperformed white students tested.

The Los Angeles Talented and Gifted program began in the 50s and was the foundation for the Enrichment Program. To this day, it is described by the Los Angeles Unified School District as "LA's BEST Afterschool Enrichment Program, which plays a vital role in the lives of unique and talented elementary school students who attend nearly 200 LAUSD schools."

In 1951, the Los Angeles Unified School District was one of two districts within California that hoped to develop an extensive educational program for those who, after extensive testing, were considered gifted. I became a part of this program. However, there was no legislation in place to provide school districts with the funding to further develop programs for gifted students. In 1956, The California Department of Education conducted a "State Study of Educational Programs" that were at that time, sponsored by the State Legislature in 1957-1960. The results of the study proved conclusively that *"special provisions made in these programs are beneficial for the gifted student and the participating pupils continue to make striking gains in achievements with both personal and social benefits."*

By 1961, AB362 had provided minimal funding excess cost reimbursement for Mentally Gifted Minors (MGM). However, there were insufficient funds to cover all school districts with the California school district. First in line for funding were not the schools in the black community. After a month, the program was canceled at my school. It was not until 1980 that the California Legislature provided specific legislation now AB1040 that adopted the federal definition of gifted.

The program was renamed Gifted and Talented Education (GATE).

The Marland Federal Report on gifted education focused on students who were/are identified as gifted and talented and who exhibit excellence and a capability for excellence far beyond that of their chronological peers. The definition expanded the categories to intellectual ability, high achievement ability, specific academic ability, leadership ability, creative ability, and visual and performing arts abilities. The Legislature deemed that each school district would determine categories for identification. It also provided 200 minutes a week of differentiated curriculum. A family friend's daughter is in this program today.

It was not until the legislature passed *AB555* in 1986. This specific legislature ensured that programs for gifted and talented students were continued and improved and provided funding for all school districts upon application and approval from the California Department of Education.

For many years, I could not explain why what was proven to be a budding mathematical genius, ability stopped as my elementary education progressed to the next mathematical stage in elementary school, fractions and suddenly I could not complete a basic fraction problem and I also could not stay in my seat. The inability to sit in a structured environment today is defined as Attention Deficit Disorder (ADD) with Hyperactivity and what neuroscientist Robert O. Becker defined in "The Body Electric: Electromagnetism and the Foundation of Life". These were some of the first studies into mind control, using children, specifically brainwave manipulation.

Although I was not labeled as such, I still was top of my class, with straight A's but I could not sit still with a "C" in Citizenship, which some would call being bored silly.

Even then, as I look back, if this was experimentation, this program was ineffective with me, and likely why it followed me all my life without a doubt. However, it is also true that so-called gifted children require variety and different educational stimulation to capture their attention.

Becker writes in the first publication version of this book in the back matter, which was later amended, describing the technique that is also known as Frequency Following Response, brainwave manipulation.

Later I would realize, as I connected the dots, that I was likely placed in the early stages of MKUltra behavioral modification, as well as efforts to suppress the growth and development of the Black community hoping to curtail young, gifted black children's success resulting from various experimentation. Many believe this is a systematic ideation and today with no color line. The game changer is awareness.

From that point on, my life felt subtly interrupted at various times, and if I had to guess today by whom, it was likely the agency that sent the young psychologist.

It is a documented fact that during that timeframe, many civilian organizations were taking part in MKUltra studies. Then and today, which included and includes many educational institutions, involved in Neuroscience along with secret military experimentation, reportedly, for example, Montauk. Widely determined to be a conspiracy theory, Montauk mocks the tales about Camp Hero being the site of secret government experiments involving mind control, time travel, wormholes, teleportation, and kids hooked up to wires in hidden underground labs. The fact is, underground military facilities are very real today.

The rumors developed in 1992. This was 11 years after the military base at Camp Hero shut down. A now widely debunked book reportedly lacking credibility called "The Montauk Project: Experiments in Time," by Preston Nichols, told of sinister, Nazi-style experiments that meddled — genetically and psychologically — with kidnapped local boys.

This program was not interested in those who tested on the national average, they focused on those who were above.

Determined to understand why innocent children would be subjected to these studies was at first a painful blow that resonates with me to this day. It is disheartening to know that human beings are thought so little of in the name of horrid scientific advancements.

As this program nudged me into research, I searched for answers and learned of vast testing in many areas. I understood what has fueled ongoing, and various types of monstrous experimentation that have developed into what is happening today, and brainwave focus remains key.

It is documented that with the use of flashing lights and pulsating tones, we can safely guide ourselves into any brainwave pattern, including those body/mind states associated with deep relaxation, meditation, hypnosis, and creating a mental and physical feeling that reduces stress and anxiety. Brainwaves can control audio-visual stimuli to alter states of consciousness is not new - "the knowledge that a flickering light can cause mysterious visual hallucinations and alterations in consciousness is something that humans have known since at least the discovery of fire".

Mind control, specifically Monarch Programming, is another aspect of experimentation reportedly with children. It is reported to be the systematic breaking of the mind through repetitive traumatic rituals that create intense mental and emotional shattering dissociation designed to condition the mind. People are groomed and prepared to be slaves since the day they're born, and are exposed to methods of hypnosis and torture that condition the subconscious mind to think and act specifically to avoid pain and punishment.

We accept that experimentation is a historical fact and a segment of Project MKULTRA, such as Project Bluebird which involved studies to divide the mind with our good friends, the Association of Psychiatry's specific studies who partnered with and have questionable relationships with Big Pharma drugs for $$$ profits.

Debates about mind control in the 20th century were focused on the roles of psychology, technology, and science in an "Age of Totalitarianism". Today, focus on the brain is inevitably linked to rapid and intense research into the mind body, and nervous system and its prolific explosion. During the 20th century, mind control became a science in the modern sense of the term, and thousands of subjects have been systematically observed, documented, and experimented on without their knowledge or consent.

Since the beginning of time, coercive persuasion by ancient Egyptians and the Knights Templar, for mass mind manipulation and torture, has been used by political and religious authorities through the Dark and Middle Ages, to the present day. The trauma-based Project Monarch, covers many areas within today's game plan, with claims of psychological operations such as Organized Stalking, high-tech energy weapon harassment, microwave "bombing" and neuro-linguistic programming (NLP).

Covertly, we have been victims of those who wish to hack into the privacy of the mind's computer, and our lives, and rewrite our very thoughts and beliefs for a controlled global population.

NLP is the study of communication and influence. If we add in the latest approaches and lessons from neuroscience, we can develop some powerful approaches to manage our minds and influence others. The influence of others is the key to what is happening today. This means that it's possible to lead some people's attention and thought processes. I can tell you from personal experience that the ones who they cannot control are primary targets marked for the most ruthless experimentation and again, in my case, a search for strategic methods of death.

These "handlers" and "programmers" desire to reshape the mind into weapons of war and assassination, using either trauma-based or electronic-based intrusions, sometimes coupled with ritual abuse, occult practices, and even mutilation and torture. Our past and present on this planet have hidden deep, dark secrets that must be exposed.

This is a fact, not only in the USA but globally, with focused experimentation on manipulating the human mind. **It is** the relentless concept that the human mind can be altered or controlled by patented technology and it has proven beyond a shadow of a doubt highly effective when combined with psychological operations.

Brainwashing is designed to reduce its subject's ability to think critically or independently, and to allow the introduction of new, unwanted thoughts and ideas into the mind, as well as change attitudes, values, and beliefs.

Behavior modification is a major school of psychological thought, that dominates many university and psychology departments; psychosurgery has been endorsed by a national commission established to investigate its dangers.

Early Monarch programming studies reveal that it disconnects the body, mind, and spirit, and after the link is broken, personal lives can be rearranged in a manner that the Handler desires with reprogramming. The reprogramming results in complete control with subconscious remnants of the trauma operating behind the scenes, affecting the person's thoughts and actions consciously with triggers, and is also associated with hypnosis studies.

An example of this type of programming can be seen in extreme religious cult groups, and various sinister occult practices designed to control the mind such as those who have been programmed for Devil worship or Satanism.

Trauma-based mind control can also impact and influence populations on a larger scale. Brain-computer Interface (BCI) and micro-miniature electronic circuits are making control of the mind through direct brain stimulation a real possibility.

9/11 is a prime example where people were so devastated and traumatized that they then became easily reprogrammed for acceptance of whatever followed. The Twin Towers and the spiritual supporting pillars of the world's societal structure were essentially demolished on 9/11, symbolically, allegorically, ritualistically, and literally.

The nationwide belief of protection within American soil as untouchable was shattered and destroyed. This type of mass trauma is not a new tactic. False flags are also used for trauma-based mind control on a larger scale because it is proven to change perceptions and result in mass compliance, for example, gun control enactment and disarming the public combined with the Globalist controlled media using repetitive mind programming messages.

You must admit that the sinister motivation and brilliant reprogramming resulting from 9/11 was masterful and diabolically effective. As we pull back the layers of deception this becomes apparent to the discerning eye.

When digging deeper, there is the revelation of how these dark controllers operate associated with the aim of global control. 9/11 was so shocking to people's systems that many just could not handle it. It overwhelmed our emotions and created fear. The result was that the American population became compliant to any retaliation and buried its head in the sand.

Full of decades of clever deception, these controllers and their tactics and overall comprehension are beyond the grasp of the average man or woman just trying to live on this planet.

And these possessed humans of evil deeds rarely even know each other. It's a hidden, and not so hidden, cult of non-feeling self-appointed elitist power who are heinous psychopaths out to subdue humanity like it's some kind of a game which to them it factually is.

The surreal significations of towers falling in the pillar of New York's St. John the Divine Cathedral – sculpted in 1997 and the Twin Pillars Archetype is powerfully occult.

Daily we are exposed subconsciously to the effect of occult (hidden) symbolism and knowledge on the human psyche. The military garners ongoing research and scientific breakthroughs using the electromagnetic spectrum. These elite psychopaths have weaponized everything, instead of using it for the betterment of humanity and our world.

Symbols are the most powerful influence today for programming the subconscious mind. The Twin Towers image is one of the most powerful.

In the early days, the accusation against Freemasonry being esoteric is two-fold. First, its members engage in esoteric practices, second, the Masons engage in esoteric study.

It was during this religious insurrection that Freemasonry was thought of as heretical or satanic. When the word esoteric in its most simple term is translated into "Hidden from Sight". Certainly, anything esoteric is not necessarily evil.

Since the dawn of civilization, two pillars have guarded the entrance of sacred and mysterious places. Whether in art or architecture, twin pillars are archetypal symbols representing an important gateway or passage toward the unknown. In Freemasonry, the pillars Boaz and Jachin represent one of the brotherhood's most recognizable symbols and most times prominently featured in Masonic art, documents, and buildings.

In mythology, Hercules reached the limits of the Mediterranean and raised two great columns upon which he inscribed Non-Plus Ultra–as this was the supposed border of the known world.

In the guise of The Great Work, the Rockefeller World Trade Center sought to embody the ideals of the modern age, the Age of Information, and the concept of World Peace through World Trade.

Trauma-Based Mind Control **and** blowing out the pillars symbolize, rightfully so, bringing down the house as in Samson in the temple of Dagon illustration at the top. In the case of the WTC Twin Towers, they represented the international financial world system, as was proudly emblazoned on the world's psyche since they were opened in 1970-72.

Dedicated to the Rockefeller brothers, high-ranking Illuminati freemasons, the iconic picture of these towers adorned more paraphernalia and received more multi-media exposure than virtually any modern architectural image

The resulting trauma created a completely suggestible populace. Hyper-state control, as a political tool, uses psychologically coercive techniques to indoctrinate subjects. The end justifies the means to form an elitist, totalitarian society. If you make a person behave the way you want, you can make that person believe what you want. Hyper-state control techniques are based on the same principles as mind control techniques studied in social control systems of cult groups and social bodies.

Creating anxiety and fear, inducing states of high suggestibility, and controlling relationships to assure loyalty and obedience are standard management techniques of the social body.

Desensitization through language abuse, propaganda, and junk information, as well as the elimination of individual ideas through repetition of chants and phrases or the inducement of dependence by introducing sports, games, or TV shows with obscure rules, are part of a large set of social styling methods. Pumping up disorientation, and susceptibility to emotional arousal, is increased by depriving the nervous system through special diets of junk food, prolonging mental and physical activity, and withholding rest and sleep.

At the occult level, the 9/11 event was a ritual, empowered, as is often the case, by performing human sacrifice. War is another such ritual, usually instigated and financed by this same dark cabal loosely labeled the Illuminati.

As the truth unfolds, **the lie is exposed, the spell is broken, and the illusion loses its power over you. Therefore, I and many, many others put our lives on the line founded on hope for today and the coming generations.**

But the secret weapon of 9/11 was this: it was riding high on the amplified occult, symbolic preparation of not just decades, seeing the "twin towers" as a symbol of world commerce and the "triumph of the human spirit", but seeing "twin pillars of society" throughout architecture and logos and literature for millennia, both conscious and subconscious, being destroyed before their eyes, over and over and over.

That these twin towers could be pre-placed and pre-determined to "blow up in our faces" with this desired effect may seem ludicrous to some.

But when your eyes get opened, it makes all the sense in the world, writes the Zen Gardner.

In these days and times, it is vitally important to stay awake, aware, and unafraid as the population control agenda progresses founded on technological advancement, human experimentation, and various methods designed to destroy the human brain turning humanity into automatons.

The foundation for mind control research continues to advance and expand, focused on:

1. **The Visual Pathways**
2. **The Auditory Pathways**
3. **Utilizing The Senses of Vision and Hearing to Affect the EEG Brainwaves**
4. **Neural Circuitry**
5. **Applications of Audio-Visual Entrainment and Brainwave Entrainment:**

AVE and BWE are techniques that use pulses of light and sound at specific frequencies to gently and safely guide the brain into various brain wave patterns. If used positively, by altering your brain wave frequencies, you can boost your mood, improve sleep patterns, sharpen your mind, and increase your level of relaxation, all with the simple push of a button! AVE also increases brain health by increasing cerebral blood flow and stimulating beneficial neurotransmitters, such as serotonin, norepinephrine, and endorphins.

6. Headache Research
7. Pain Research
8. Relaxation Research

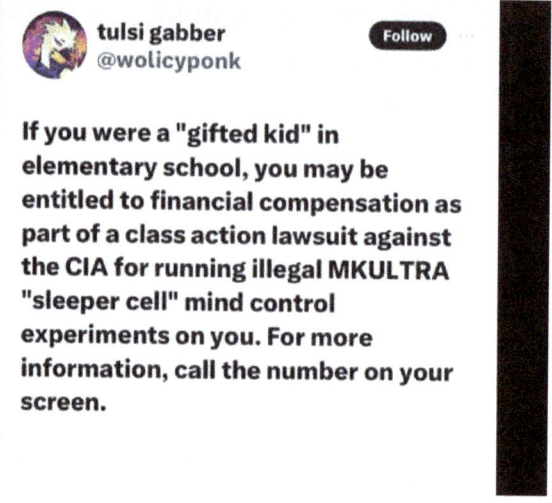

Twitter "X" Post

CHAPTER 3

The Denizens of the Abyss Call Their Names

He had hundreds of monsters inside him wearing his face as a mask, screaming and trying to tear him apart and take his place. He always fought furiously to hold them back and it created an unending chaos inside him. Eventually, in the end, he lost all his strength and battles. He was dragged down into the abyss. He cried and fought hard to find his way back home. To get out from there again and to be himself. But among all these masks, the real he was lost forever. He never made it back again, and he was not himself anymore." ~ Akshay Vasu, "The Abandoned Paradise: Unraveling the beauty of untouched thoughts and dreams."

One definition of evil defines a negative intellectual being in a positive human body. The intellect connection stems from the awareness and knowing. In psychiatry, this is called a psychosomatic characteristic caused or aggravated by a mental factor such as internal conflict. Evil is the fictitious personality out of sync with the true self. The question is, "Is any person doing evil an actual metaphorical devil? The earliest recorded definition of 'angel' in the Oxford English Dictionary (OED) is documented in the *Lindisfarne Gospels* in

950AD, along with 'devil' which has its first OED citation 150 years earlier and both have undergone little change in the literal meaning since. The Mapping Metaphor resource demonstrates that 'devil' is the polar opposite of 'angel'.

The fact is anyone can wear the jacket of evil by desire or choice. Evil manifestations exist in the universe because of desire lingering until it has found true existence. Humans give evil a meaning through our intellectual evolution derive from the intellectual mystery and hope for answers of how and why demanded by reality.

Existence has two dimensions within the polar state of Earth. This duality is a rule of existence, along with consciousness, and nothing can exist without the rule of duality and both. We exist as man and woman, experiencing negative and positive, light and darkness, yes and no, love and hate, God and Devil, good or bad, etc.

Evil in and of itself is lack of empathy, or concern for others. From this point of view evil is "the opposite of compassion" with compassion regarded as the highest form of intelligence. This is because intelligence from a compassionate point of view requires considering others and greater understanding.

The lower the level on the ladder of consciousness evolution, the lower the interconnectedness of its elements and the more primitive the Soul. The lower the connectivity between neural networks in the brain, the more dissociative states individuals experience resulting in decreased insight while the lower states of consciousness increase to narcissistic ego tendencies culminating in a malignant narcissism, heinous sociopathy or psychopathy. Empathy is vital for a healthy perspective in life and without it, there is an inability to connect to the environment in a meaningful way leading to a primitive level of functionality observed in lower organisms. For example, when the category of animals and habitats is transferred to 'devil' we use 'beast' in the sense of "the animal nature (in humans)"

Within the mentality of the beast lies the undeveloped brains of socio and psychopathic individuals as recognized traits determined by their thinking and primitive predatory survival mechanisms akin to those of chameleons.

Evil lives within as an organic instinct. It rears its ugly head as an inability to treat others as you would like to be treated and is warped and corrupted by perversions of other instincts, lies, insecurity, doubt, and mainly fear. Fear triggers greed (I don't have enough or I'll never have enough), hatred (this person threatens me in some way), or low self-esteem (No one loves me or ever will). These are the building blocks of what we call "evil." The once shining spark of goodness becomes diminished over time, slowly succumbing to the darkness that infects them. The fact is, the world is full of evil and cruel people who are ready to devour anyone to feed their lower-level consciousness. They take and take then they throw them out when they no longer need them and after successful destruction they move on to someone else.

"Evil" is viewed as just another word for high entropy consciousness or the combination of chaos and the polar state of the gradual decline into disorder. Entropy is the term used to describe the progression of a system from order to disorder. Picture an egg: when it's all perfectly separated into yolk and white, it has low entropy, but when you scramble it, it has high entropy and it's the most disordered it can be. In psychology, this is also called the Cluster B personalities. Cluster personality disorders are specific types of mental health conditions. They can lead to consistent, long-term, and unhealthy patterns of thinking, feeling, and behaving.

Upon closer inspection of the individual, and the cruelty they inflict consciously is the revelation that they lack the specificity needed for rational understanding. In essence, evil is a thought, word, or deed, that is completely aimed at benefiting oneself and oneself alone while satisfying horrific needs. Why does evil exist? Some argue that God allows evil because evil produces additional goods that would not be possible in a morally pure world, such as the courage gained by the victim of the lower level of consciousness.

However, make no mistake about it, evil when dealing with those possessed and blinded by evil, is powerful, and within the duality of polarity, it can be equally as powerful as good because of this, it is understandable what Friedrich Nietzsche wrote, "He who fights with monsters might take care lest he thereby become a monster. And if you gaze for long into an abyss, the abyss gazes also into you."

Mind Control and the Destruction of the Sacred Self

Dating back to the Cold War and even further, historically, has evolved into a high-tech "Crime of the Century." Since the dawn of time, those who deem themselves the controllers of humanity have embarked on an exhaustive, relentless search, to continued control over the masses and to retain their powerful influence and subjugation by any means necessary. This quest, yesterday, today, and tomorrow has resulted in an ongoing antediluvian desire for ongoing mass human experimentation for social and mass population control ideation as efforts designed to subdue the population, create distractions, chaos, and create instability. One of the twenty-first century's greatest violations of Human Rights, unspoken, is the proliferation of mind control technologies and the unspeakable heinous use and abuse combined with high-tech torture and slow kill deaths. Today, as the veil lifts, thousands of human guinea pigs, have taken a stand raising their voices in awareness and protest of this reality.

The prolific use of mass mind control technologies, which use electromagnetic waves to permeate the brain and nervous systems are brilliantly patented to subvert an individual's sense of control over their mind, body, and spirit, encompassing and simultaneously distorted thinking, modified behavior, emotions, and decision making. These weapons remain in ongoing research, testing, and development programs by the military, united with DOD contractors, Lockheed, Raytheon, Northrop Grumman, and a host of others which demand large-scale experimentation that in the world has become a monstrous system of horror. Yet, the desire to expose mind control technologies and their torturous abuses, today has grown by the determined efforts of people all over the world to urge governments to halt these horrific violations of Constitutional, Civil, and Basic Human Rights.

The battle is likened to David and Goliath, founded on the belief by Targets that full disclosure is vital and relief will not be achieved unless those unaware of mass human experimentation are aware and become just as outraged as those who are being victimized. The horrific use of these patented advancements remains a determination of ongoing and logical coverups through unbelievably inhumane, monstrous types of individuals. Neuroscience uses Artificial Intelligence, Hive Mind, electronic microchip implants, nanotechnologies, microwaves, and/or electromagnetic waves to subvert an individual's sense of control over themselves by attacking the brain and nervous system leaving surreal destruction in its path.

The development of high-tech methods of covert destruction using bioweapons and mind-invasive advanced, psychotronic, and psychophysical technology has a long history. Research during World War II and after (Cold War) became the foundation for heinous human studies including the search for powerful drugs, tested on concentration camp inmates such as meth and cocaine-based, with both believed to be wonder drugs. Hitler used himself and hoped to enhance the performance of German troops and today's use to strategically subdue individuals and groups. communities and large populations, ruthlessly or "in the so-called, "name of science." Using drugs, which essentially are biological warfare continues with documented proof of bioweapon street drug infiltration from poverty-stricken communities to the affluent within a long-held belief that also lower-income, less educated, voiceless people specifically are easier to control.

Experimentation runs the full gamut of methods from mind control to Voice to Skull technologies with many diagnosed falsely as schizophrenic on a massive scale. As documented in this book series overall, artificial microwave voice-to-skull transmission was successfully demonstrated by researcher Dr. Joseph Sharp in 1973, announced at a seminar at the University of Utah in 1974, and in the journal "*American*

Psychologist" in the March 1975 issue, the article was titled *"Microwaves and Behavior"* by Dr. Don Justesen (1975). This was merely the initial official documentation of technology used for beamed communication experimentation, for many years to come.

Highly effective technologies today include mind reading, forced memories, dream manipulation, auditory and visual hallucinations, and various types of horrific experimentation and for some much worse. Exposure is vital!

Ongoing research into electromagnetic spectrum weapons has been secretly carried out in the US and Russia since the fifties. Plans to introduce these super-weapons were announced quietly in March 2012 by Russian Defense Minister Anatoly Serdyukov, fulfilling a little-noticed election campaign pledge by President-elect Putin. Mr. Serdyukov said, "The development of weaponry based on new physics principles - Direct Energy weapons, Geophysical weapons, Wave-energy weapons, Genetic weapons, Psychotronic weapons, and so on - is part of the state arms procurement program for 2011-2020." The USA in this modernized race for this type of weaponry will not be left behind as also a leading force for development.

There are many reported stories of some poor souls who have claimed they were provoked, influenced, tortured, and harassed by remote voice-to-skull beamed technology producing and describing the hearing voices effect characteristic of electromagnetic mind control technologies. Most of the public, including the media, have strategically labeled anyone revealing this truth as conspiracy theorists.

In the days and weeks before authorities say he shot three and injured one at the Florida State University library and was then gunned down by Tallahassee police, Myron May, the Assistant D.A. Las Cruces, NM, posted a video about mind control, Directed Energy Weapon attacks to his heart, "Organized Stalking" is another aspect of mass human experimentation key to this program.

May also posted a video of former professional wrestler and Minnesota Governor Jesse Ventura interviewing a man who claims to have created technology, Dr. Robert Duncan, that allows the federal government to control people's minds. This was part of the Jesse Ventura, TruTV "Conspiracy Theory" episode entitled "Brain Invaders" which first aired December 17, 2012. However, the fact is, by typing into the Google search engine "people who said that God or Satan told them to kill" you are inundated with news reports across the USA and globally subsequent atrocities after given verbal instructions inside the person's head to kill.

Jimmy Shao of Sacramento, California, was arrested for calling 911 more than one hundred times in one month in May 2013. Shao said he would not stop until Congress investigates the shadow government who use satellites to control his mind and body.

Jared Loughner, 23, who accused of shooting a US congresswoman in Arizona and killing six others on January 8, 2011, claimed that he was being mind-controlled. Fayette woman Angela Modispaw claimed she heard voices telling her to kill her mother in 2009.

Another suspected victim, Honduras' fallen leader told *The Miami Herald*, he was being subjected to mind-altering gas and radiation - and that "Israeli mercenaries" are planning to assassinate him.

Ronald Morgan, 18, a teenage high school dropout who told investigators he was acting on God's orders confessed to beating his father to death with a baseball bat on May 27, 2001. Morgan said God had told him in a dream to kill his parents.

Michael Robert Lawrence, accused of murdering a vacuum cleaner salesman in Waialua, said he was on a "mission" to kill people and chop up their bodies after voices commanded him to do so, a psychiatrist testified on April 3, 2001.

Richard Scott Baumhammers, 34, was arrested Friday, April 28, 2000, following a shooting rampage that left five dead and one seriously injured in the Pittsburgh, Pennsylvania area. Baumhammers told a psychiatrist that he could hear people talking about him.

Solomon, 15, allegedly opened fire on other students at Heritage High School on August 10, 1999. He heard voices telling him to do strange things, but they were robotic voices, not human voices.

Tami Stainfield is a woman with evidence that proves she and others are victims of predictive analytics robotics and human logistics. She claimed that "we are tortured, hostages, and slaves to a network of technology void from identification and protection." She filed as a "no party" candidate in 2012 for the Presidency of the United States.

The majority of victims' stories had been ignored by the media, under media Gag Orders, to insure an unaware public as the death toll explodes which includes Mass Shooters report being mentally controlled into creating atrocities.

***

The United States of America's Army's "Military Thesaurus" defines "Voice to skull" (V2K) as a misnomer meaning non-lethal.

"Voice to Skull devices are reported to be nonlethal weapons. Yet with relentless use on human guinea pigs combined with microwave weapon attack deterioration of a target's health the result is high-tech destruction to a person's life is diabolical, gradual and intentional. The psychophysical breakdown of the person targeted by these weapons is anything but nonlethal. Neuro-electromagnetic weapons, systems and devices use microwave transmission of sound into the skull of persons or animals by way of pulse-modulated microwave radiation.

Most cannot imagine a weapon that creates sound that only you can hear. Yet these weapons are patented and in full use and have been for many years veiled by ruthless secrecy. The military are key players with the US Air Force, Navy, etc., having documented experimentation with microwaves that create sounds in people's head used as a tool of Psychological Warfare. American Inventors such American Technologies. Hypersonic Sound System. Woody Norris, can isolate sounds to specific targets in a crowd as does the Audio Spotlight Sonic Weapons, identical to what was reported at the Cuban Embassy. "Havana Syndrome" held the spotlight in the news for a period of time until interest was lost. The symptoms range in severity from pain and ringing in the ears to cognitive difficulties and were first reported in 2016 by U.S. and Canadian embassy staff in Havana, Cuba.

Microwave Attacks on Americans (aired 2/20/22)

The Pentagon's scientific arm for the advancements of various types of weapons for military use, the Defense Advanced Research Projects Agency (DARPA) with their new "Sonic Projector" program, hopes to expand this same technology. These weapons have now been filtered to all levels of law enforcement and are being used covertly.

The goal of the Sonic Projector program is to provide military Special Forces and again, law enforcement with a method of surreptitious audio communication directly into the brain or around the head at distances over 1 km. Sonic Projector technology is based on the non-linear interaction of sound in air, translating an ultrasonic signal into audible sound. For decades, anti-personnel directed energy devices that cause targets to hear voices (and other sounds) that other people near the targeted individuals cannot hear have been a major part of ongoing destructive experimentation and obviously in the hands of the mentally disturbed, who are the real crazy ones. The beamed voice of the official operator can be loud enough to be heard only by the target by irradiating around the head or so quiet that it has a subliminal influence effect. Herein lies the danger, when an unaware person thinks that a suggestion is their own and as a result acts on it thinking instructions from God or Satan.

The technology can penetrate any surface, solid brick, and stone walls effortlessly in the 24/7 tracking and monitoring paradigm. The fact is, the experience of Synthetic Telepathy or Artificial Telepathy, Remote Neural Monitoring, the Frey Effect, Microwave Radio Frequency Hearing, the Microwave Auditory Effect, etc., is not that extraordinary. It's as simple as receiving a cell phone call in one's head. Most of the technology involved is identical to cell phone technology, which uses microwaves for communication systems.

Satellites today, combined with the explosions of official, psychotronic drones, 30,000 approved for U.S. Skies by 2020 link the sender and the receiver.

A computer "multiplexer" routes the voice signal of the sender through microwave towers to a very specific defined location or cell. The "receiver" the target is located and tracked with pinpoint accuracy, to within a few feet of their location. The receiver of the communication is not a cell phone.

It is the human brain. Out of nowhere, the voice of the official operator suddenly booms in the target's head and it cannot be stopped as reported by Timothy L Thomas (The U.S. Army War College Quarterly) who reported years ago, "The Mind has no Firewall, therefore is susceptible to all forms of non-detectable intrusions." He further states:

The human body, much like a computer, contains myriad data processors. They include, but are not limited to, the chemical-electrical activity of the brain, heart, and peripheral nervous system, the signals sent from the cortex region of the brain to other parts of our body, the tiny hair cells in the inner ear that process auditory signals, and the light-sensitive retina and cornea of the eye that process visual activity.

We are on the threshold of an era in which these data processors of the human body may be manipulated or debilitated. Examples of unplanned attacks on the body's data-processing capability are well-documented. Strobe lights have caused epileptic seizures. Recently in Japan, children watching television cartoons were subjected to pulsating lights that caused seizures in some and made others very sick.

If your thoughts were hacked, you wouldn't even know that thoughts that materialized in your mind weren't your own nor could you separate yourself from the subliminal influence of making suggestions and you think it is your own idea brilliantly in your voice characteristic of this capability. Like a demented phone caller, an evil person at the helm with this technology, in uniform, in hand can call around the clock beaming negativity, degradation, and relentless harassment until the person breaks, or is programmed, which is deemed a successful operation that can culminate in death.

Perhaps the most horrendous use today remains the use shown above by puppets, and below that it has influenced casualties, and ongoing reports of people reporting that a voice, make no mistake about it a human voice, which they mistook for God or Satan telling them to do something deadly to even loved ones and horrendously reported by men, women and children:

Jason Dalton said that when he logged onto [Uber's app], it started making him be like a puppet," the police documents said. He claimed that the devil's head "would give you an assignment, and it would literally take over your whole body."

Arman Torosyan, 23, killed his parents – 64-year-old Khachik Torosyan and 57-year-old Marietta Torosyan in their apartment in Sevan, as he says, "fulfilling the commandment of Jehovah."

Thomas Hammer, a San Clemente teacher, reported that a "Higher Power" told him to attack a kid on a skateboard. "When I stepped in, I felt compelled by a higher power," Hammer told the Orange County Register. "Honestly, have you ever been grabbed by the Lord in a way you never thought you would or you could? That's exactly what I'm testifying to, and I'm not speaking in hyperbole. I'm speaking right from the heart."

Police in Richland Township, PA, a 26-year-old man, Levi Staver Daniel, killed his grandmother while she ate breakfast, then claimed the Archangel Michael told him to do it

Teen claimed Satan told him to kill his parents. On March 25, 2008, the teen attacked his parents with a knife at their Baulkham Hills home, killing his father and wounding his mother, saying Satan told him to dial 5455 and kill. Those numbers produce the word "kill" on a cell phone keypad, reported the 16-year-old boy in Sydney who said the devil told him to kill his father and wound his mother.

On Monday, 27-year-old Ayanie Hasan Ali casually strolled into a recruiting center in Toronto and began stabbing two members of the Canadian military while shouting "Allah told me to do this, Allah told me to come here and kill people." Ali is a Muslim and a citizen of Canada. Police have leveled multiple charges against the attacker, including attempted murder.

The nanny accused of beheading a four-year-old girl in Moscow and waving her severed head outside a Metro station told journalists before a court hearing that 'Allah ordered' her to murder the child.

A Texas woman who stoned two of her children to death and seriously injured a third on Mother's Day told psychiatrists she was driven to kill by a message from God and that she was sure they would rise again from the dead. Files said Dianna Laney believed God had told her the world was going to end and "she had to get her house in order," which included killing her children.

Cody Edmund Dixon, a Texas man who allegedly killed his girlfriend and their 9-month baby, said God instructed him to do it.

Mom, Marie J. Chishahayo accused of letting her son kill her daughter, says God told him to do it.

Tammi Estep, a woman who allegedly stabbed her husband, said she did it after, "Jesus and Mary told me to kill him because he is Satan's spawn!" according to a police report. She allegedly added, however, that "Jesus" had told her to drown her son. Police also say she claimed her son had "become stuck to the ground."

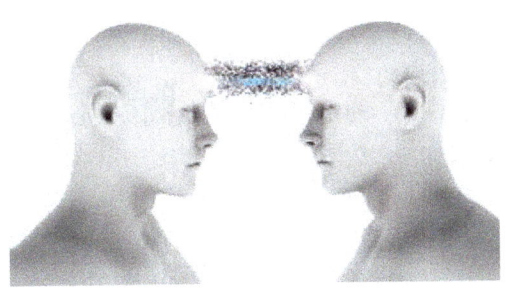

Extrasensory perception (ESP) refers to information that is perceived outside of the five senses. This includes phenomena such as telepathy, clairvoyance, and knowledge of future events.

Since these phenomena cannot be overtly seen or measured, they are often regarded as unbelievable. However, recent research explores the biological mechanisms behind such phenomena.

Mirror Neurons: Telepathy refers to communication outside of the known senses. Many studies have shown that we can "read" other people's minds because we have neurons that act as automatic mirrors. We can grasp the intentions and emotions of others automatically.

In 2007, psychology professor Gregor Domes and his colleagues found evidence that the ability to interpret subtle social cues can be enhanced by oxytocin, a hormone that increases trust and social approach behavior. It is not such a stretch to imagine that we can pick up the emotions and intentions of others around us, but can this be done when long distances separate people?

Long-distance communication: One study conducted in 2014 study by psychiatrist Carles Grau and his colleagues found that brain-to-brain communication via the Internet is possible.

While this is alarming, the fact is that official use today is by a patent technology capability and has been patented for decades! God nor Satan speaks in a human voice, but heinous official socio and psychopaths do. This technology with worldwide complaints is in use all over the world.

Victoria Soliz allegedly tries to drown her son in a puddle because Jesus told her to.

Accused Murderer Says God Made Him Kill - The community of Tyler, TX, is reeling in the aftermath of a gruesome murder that would make Jeffrey Dahmer proud. 25-year-old Christopher Lee McCuin has been arraigned for allegedly killing and mutilating his 21-year-old girlfriend, Jana Shearer.

Dora Tejada told investigators God told her to push a rose in the girl's throat to exorcise the devil. Dora Tejada, 28, was found not guilty by a judge today in Nantucket Superior Court and instead committed to a state mental health institution for six months of observation and treatment.

Hearing commands from God in this context is characterized by almost all courts and commentators as a sign of mental illness. Yet nearly a third of Americans believe God speaks directly to them in personal revelations or prayer, some seeing an image and/or hearing an actual voice and words, others experiencing a "thought-insertion" from "outside" themselves. The facts are similar to stories or examples from the Western cultural record, in which we rarely dismiss the God-hearing protagonists as insane.

To mention only a few: Jesus, Abraham, Joshua, Moses, Joan of Arc, John Brown, and various historical Popes are usually not presumed to be insane, but sane, honest, and, for the faithful, true recipients of God's commands. Many of these revered figures have even committed homicides purportedly at God's behest. Yet, God nor Satan is human and does not speak in a human voice. And sadly, if a hologram is used to convince people that Jesus is coming or has arrived, in the sky, those who believe this as part of their worshipped religious doctrine may be easily duped.

During the summer disturbances in Washington, D.C., a top local military police officer asked the D.C. National Guard about deploying two military systems that seemed to come out of science fiction. One, the Active Denial System (ADS), makes the target's skin feel like it's on fire. The Long-Range Acoustic Device (LRAD), directs intense sound in a narrow cone. The sound is so clear and so powerful that it was nicknamed "the voice of God." I encountered both systems, one in Quantico, Virginia, and the other in Falluja, Iraq. Here's what I saw.

***

In recognition of patented technological capabilities, today, there must be an awareness of these advancements, again in research, TESTING, ongoing development programs for DECADES, and factual disclosure of the capabilities of beamed verbal communication technology used by anyone. The hope is to save lives with this technology in the hands of ruthless official personnel devoid of human compassion. Within certain circles, awareness and use are well known, especially by those we are programmed to trust as heroes.

Using these weapons demands full disclosure:

Can we logically expect advanced, ultra-secret technology that is not only space-based but land and sea, etc., weaponized and connected, to an intricate electromagnetic frequency system used for mass surveillance, and mass control, that can lock on and track and monitor human beings, simultaneously, with a pinpoint accuracy that can focus on manipulation of brainwave frequencies to be publicized? Again, brainwave manipulation is the key to mind control and behavior modification. Or can we expect exposure to technology that uses various types of energy fields to tamper with our natural shield for global control to be publicized? No!

The patents deriving from Bernard J. Eastlund's work provide the ability to put unprecedented amounts of power in the Earth's atmosphere at strategic locations and to maintain the power injection level, particularly if random pulsing is employed, in a manner far more precise and better controlled than accomplished by the prior art, the detonation of nuclear devices at various yields and various altitudes." (ref High-Frequency Active Auroral Research Project, HAARP).

"Harnessing neuroscience to military capability, the advanced technology today results from decades and ongoing research and experimentation, most particularly in the Soviet Union and the United States. (Welsh, 1997, 2000)

The fact is, these technologies should no longer enjoy the secrecy they have held for decades and many highly credible people agree who are stepping forth with a unified Earth-shattering voice. As they do, the evildoers at the helm realize that the truth is their nemesis. They are those who operate at the lowest level of consciousness, homebound for the abyss, who celebrate their deeds and know their names, waiting patiently for their return with their reservations on hold.

I pitied them, as I suffered attacks to my right leg day to day remembering what Dean F. Wilson wrote, "No one, not even the demons of the Regime, could survive at the immense pressures of the abyss, and even if they could, they would likely drown before they set eyes upon the gaping maws of the denizens of the deep." ~ Dean F. Wilson, Lifemaker.

I also recognized that anyone who wishes them harm, no matter how cruel and evil they are sitting at the helm of this technology in this hideous program, that if you do, the abyss would truly gaze back at you, and you could become them acting within a lowered frequency which should be frightening.

The remedy, and it is imperative, is to vibrate at the highest frequency of love while fighting the good fight!

They create their destiny and don't need your help. Save your thoughts to create your reality, founded on your inner strength and endurance, becoming a conqueror! Thank God for the awareness. Some do not have a clue they are being monstrously used.

Understand one thing: this technology, as shown below in the References, is very real and NO JOKE!

Be aware that for this type of cruelty to strive without consciousness is because of the evil that has engulfed these people and it is heinously ugly.

Within the despicable coverup hiding, the truth is founded on lies, destruction, deceit, deception, death, charades, and everything hideous you can imagine, fueled by what appears to be a thirst and destructive personal psychosis of those involved. Their programming is so intense that they call evil good. There may not be help for them, but awareness can help to save lives from this evil of men who are clean-shaven, suited, and in uniform.

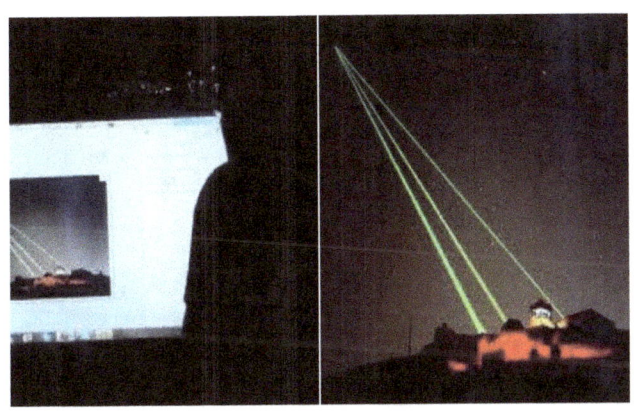

CHAPTER 4

Falling into the Abyss Has Never Been the Fate of Those With Wings

"Evil dwells and sits on the throne.
Wherever the abyss of chaos calls home
Evil is the destructive loveless force
That tears lives and souls
apart as its main course
Evil will manipulate coerce and seduce
It is the deranged self's obsessed mind
Evil's every thought and deed curse the
Divine having 'no line' that divides
The truth is destroyed from the lies
used to captivate and brainwashed
Vulnerable minds. The abyss agenda as
a vessel they try to apply until the
day they die"

~ Anonymous

Today, the mere thought of the ability to change the human mind is ludicrous to the vast majority. No one can accept the reality of an effortless capability, to control another person's thinking, nervous system, or body, with everyone believing themselves to be the masters of their destiny, which is simply not true. The belief that the human mind can be manipulated and influenced by outside sources is commonly the plot in horror movies. This is although history documents that every civilization from antiquity to the present day has had some form of mass mind control and used various methods and techniques to ensure control over their flock demanding mass population control. Today, the ideation emerges as the focus on the brain itself, with the seriousness of this capability, something that cannot be denied or written off as harmless.

Brain entrainment is also described as being controlled by thoughts that materialize within the subconscious and are inserted artificially. History throughout the ages records many tactics used in every civilization, documenting specific types of successful efforts used for control over others. For a greater understanding, let's look back to the role of hypnosis, one of the first practices for mind control, and specifically how states of consciousness were used for healing and greater understanding in earlier times.

Over 5,000 years ago, the Temple of Imhotep was designated as a healing center. The process entailed herbs and the rhythmic recitation of prayers. The individual seeking a cure was led to a special darkened chamber to sleep and await a dream revealing a cure in what were called "sleep temples" first in Egypt and then in ancient Greece.

Using hypnotic states was also exemplified in the ancient practice of oracles. Like Sleep Temples in Egypt and Greece, individual expectation and overload were essential ingredients for both the oracle and the subject. Today, we now understand the divine answers and feelings of reassurance experienced by ancient peoples as being the product of sensory overload, expectation, and direct suggestion, but that does not diminish the physical and emotional healing that took place because of it.

A review of the Egyptian Book of the Dead also provides many occult methods to manipulate the mind and methods include torture, and intimidation, designed to break targets down through fear and pain, creating unforgettable trauma that opens to door for complete control of a person. The person becomes controlled because of the fear of being subjected to such treatment (trauma-based mind control) resulting in total submission. Drugs have long played their role, similar to herbal rituals for mind control because the human mind is more pliable than it would be in its normal state.

From generation to generation, these basic techniques combined with various types of intense experimentation have been perfected, and resulted in designs for application, on a massive scale, technologically. Not only today is this happening through Artificial Intelligence (AI) and Brain-Computer Interface (BCI) technologies but also as simple as the use of the repetitive nature of mainstream news, movies, music videos, advertisements, and television shows that relentlessly reinforce a frame of thought. These techniques combined result in the evolution of humanity to a state of subtle, perfected mass mind control.

A look back in history reveals the origins and foundation for hypnosis use and covert nonconsensual experimentation today:

The tactic of mind control, and bearing false witness use of others, to destroy a person by methods such as hypnosis, is also a historical fact brought into the open with the Cardinal Jozsef Mindszenty's trial in the 1940s. Cardinal Mindszenty was forced to confess, although he was not, that he was a spy in a Hungarian court and was convicted of treason and sentenced to life imprisonment.

Convinced that the cardinal was actually "the center of the counter-revolutionary forces in Hungary," the communist authorities wanted him. The immediate issue was an order that Hungary's 4,813 Catholic schools be nationalized. The cardinal had driven from village to village, urging people to ignore the communist lies and refuse to give up their schools and their land. The police responded by seizing his sound truck and portable generator. At his order, church bells tolled.

Although the publication of his last pastoral letter was banned, one copy made it to "The Voice of America" radio broadcast, and the communists shouted "subversive activity."

The November 1948 letter ended:

"I stand for God, for the Church, and for Hungary. . . . Compared with the sufferings of my people, my own fate is of no importance. I do not accuse my accusers. ...I pray for those who, in the words of Our Lord, 'know not what they do.' I forgive them from the bottom of my heart."

Explaining that neither he nor the Church had provoked the enmity of the Hungarian government, he wrote in an open letter in December, "Communism is an atheistic ideology: hence by its very nature it is opposed to the spirit of the Church."

When the Cardinal admitted to being a spy, the effectiveness of mind control was applauded. We never forget that the objective of mind control is to change a person from their normal self.

***

Hive Mind technology and the patented Silent Sound Spread Spectrum, patented many years ago, today reveal a sophistication derived from the wealth of historical knowledge, from which the most advanced methods to manipulate human behavior ever compiled in our day and age have emerged efficiently.

True, the Silent Sound Spread Spectrum is often used on alleged conspiracy websites in connection with brainwashing and mind control. However, it is a factual, patented technology, patented at the United States Patent and Trademark Office, as US5159703A - Silent Subliminal Presentation System.

The silent sound subliminal mind control technology today is a space-based system that produces a steady tone near the high end of the hearing range of about 15,000 Hertz, which is the same one the hypnotist's voice uses ranging from 300 Hertz to 4,000 Hertz. A steady tone inaudibly injected at 15,000 Hertz, accompanied by the hypnotists or today's official personnel at the helm's voice ranging from 300 Hertz to 4,000 Hertz, that passes into a frequency modulator.

The operator controls the frequency modulator as a steady tone coupled with their voice, which is like tinnitus ringing in the ears, but with the hypnosis agenda embedded. As a result, the FM frequency modulated controls the timing of the transmitter's pulses. The transmitter itself emits one short pulse of a microwave signal at a frequency to which the human brain is sensitive, then the brain converts the train of microwave pulses back to an audible voice inside the brain using timing. Each microwave pulse is controlled by each downslope crossing of the voice. The total output of the system goes into your brain and there is no conscious defense possible against this silent sound of hypnosis and hypnotic suggestion. There have been many patents issued for this technology. The patent and its patented process are real. It's not imaginary or mental illness. This technology was the tip of the iceberg and evolved into what is being reported by people within the United States and globally.

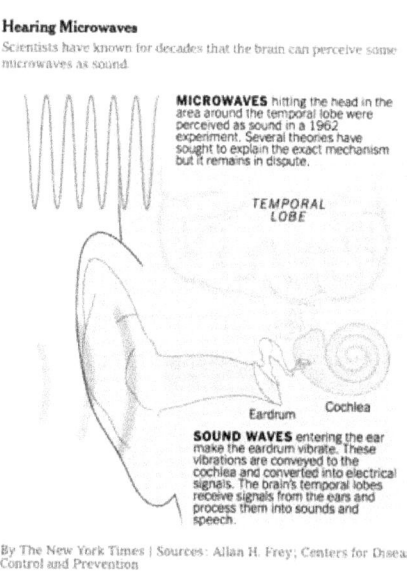

By The New York Times | Sources: Allan H. Frey; Centers for Disease Control and Prevention

Who could carry off something like this on a massive scale? The fact is the military for one. The USAF has focused on advancement of various types of this technology, and it now in the hands of Federal, state and local police departments.

As shown in the article excerpt following regarding Artificial Telepathy, entitled "Focus On U.S. Air Force Space Command" it also reveals patents.

[Excerpt]

In this post we will begin a closer examination of U.S. Air Force mind control operations. Why Air Force? And why start with a focus on the Air Force Space Command? For several reasons:

- The U.S. Air Force imported dozens of Luftwaffe scientists during the 1940s and 1950s under Operation Paperclip
- Many of these Nazi scientists worked at the School of Aviation Medicine near Randolph Air Force Base, San Antonio, TX during the 1950s, conducting (among other things) pilot stress tests, experiments in radiobiology and Human Radiation Experiments

- Hubertus Strughold, a Nazi doctor who worked at SAM, went on to work for NASA, developing remote telemetry sensors that could monitor the vital signs of astronauts travelling to the Moon --a mind-boggling feat of precision that suggests remote neural monitoring is indeed quite possible

- Today, SAM and the Human Effects Center of Excellence (HECOE) at Brooks Air Force Base help the Air Force Research Laboratory test its next generation of non-lethal weapons, including lasers, masers and microwave weapons used for crowd control

- The Air Force Research Laboratory (AFRL) has done extensive research into microwave hearing and voice-to-skull devices, according to a Washington Post article on voice hearers at the following link:

- https://microwavenews.com/news-center/mind-games

- AFRL is also one of the main sponsors of HAARP, a huge antenna farm near Fairbanks, AK, that is reportedly broadcasting in the 8-12 MHz (brainwave) range, and capable of delivering signals to almost any point on the globe

- HAARP is technically part of the Air Force's Ballistic Missile Early Warning System, designed to detect missile launches coming from Russia

- HAARP therefore falls under the direct command and control of NORAD, the U.S. Northern Command (NORTHCOM) and the U.S. Space Command in Colorado Springs, CO

- US Space Command also operates spy satellites for several other agencies of the Department of Defense, including the NSA, NASA, the National Reconnaissance Office, and the Defense Intelligence Agency.

- Artificial Telepathy.blogspot.com" Reference Link

The agenda for mass mind control does not stop with mass experimentation on the civilian population but also includes efforts to keep those at the helm of this technology, military personnel, and all others mentally controlled, as shown in another excerpt below.

The Pentagon's blue-sky research arm wants to trick out troops' brains, from the areas that regulate alertness and cognition to pain treatment and psychiatric well-being. And the scientists want to do it all from the outside in — with a gadget installed inside the troops' helmets. "Remote Control of Brain Activity Using Ultrasound," the Defense

The foundation for the techniques used today originated from a compilation of rituals, heavily studied by governments, militaries, and today's secret societies with members operating in intel agencies and military leadership and practices that describe methods of torture and intimidation (to create trauma) the use of potions (drugs) and the casting of spells (hypnotism), and ancient events ascribed to black magic, sorcery and demon possession (where the victim is animated by an outside force). These techniques are ancestors of modernized mind control and weaponization of brilliant electromagnetic spectrum-derived biotechnology. Potions can today be likened to both illegal and legal drug experimentation and real-time monitoring for the advancement and search of psychotronic and combined bio-weapons.

As we know, Project MKULTRA was an illegal human experimentation program designed and undertaken by the U.S. Central Intelligence Agency (CIA) and intended to develop procedures and identify drugs that could be used during interrogations to weaken people and force confessions through brainwashing and psychological torture.

The process of mind control has always focused on changing the personality and it is so intense that the target becomes a living puppet, a mere human robot, a shell of his true self, and this is accomplished without evidence of psychophysical tampering visible to the person.

A flashback to the early 60s is a reminder of one of the most highly publicized reported puppets of all time. The man convicted of assassinating Robert F. Kennedy.

There have always been fall guys used for many high-profile assassinations, ideal candidates for prostitution, today reinvented as Human Trafficking, or brainwashing for billion-dollar sexual, pornographic movie production.

Indoctrinated, unconscious automatons have been proven historically to use hypnosis to carry out instructions. The curious thing about Kennedy's assassination is that Sirhan Sirhan many believe was a scapegoat, and this was reported to the psychologist, Dr. Eduard Simson-Kallas, who met with him after sentencing to San Quentin in 1969. This doctor was one of the first to associate mind control and hypnosis and publicizing the possibility when interviewed by the San Francisco Examiner, stating that Sirhan was hypnotically programmed.

Hypnos-programmed agents, unconscious automatons blindly acting out other's instructions, have been popular subjects of speculation.

The fact is, your mind is being **controlled by distant strangers who don't have your best interests at heart.**

What struck Dr. Simson-Kallas as unusual and aroused his suspicion after he spoke with Sirhan was that he could not explain his crime in a vivid language. The doctor stated that he seemed to be programmed and "reciting from a book," and "The curious thing was that he didn't have any details.

A psychologist always looks for details. If a person is involved in an actual situation, there are details." As trust built between the two men, Sirhan trusted Simson-Kallas. Believing he was hypnotized, he asked to be hypnotized. "I don't know what happened," Sirhan said. "I know I was there. They tell me I killed Kennedy," Sirhan continued. "I don't remember what exactly I did, but I know I wasn't myself. I remember there was a girl who wanted coffee... So, I gave her my cup and poured one for myself. That's the last I can remember until I was choked and manhandled by the crowd."

Most are familiar with Project MK ULTRA, however there were several other "Projects" during this era.

There have been many allegations that Manchurian Candidates exist. Within the last two decades, the idea has been seriously presented as the explanation for political assassinations in the United States and elsewhere, including mass shootings for gun control enactment to disarm the rebellious public.

Hypnos-programmed agents, unconscious automatons blindly acting out other's instructions, have been popular subjects of speculation.

Infamous James Holmes was among the elite of Neuroscience before the Aurora theater massacre.

As James Holmes sat bewildered in the Courtroom of a Colorado judge who told him he faced murder charges, James Eagan Holmes was supposed to give a presentation on a topic so complex that most people would barely understand its meaning or relevance. He was prepared to discuss "microRNA biomarkers" with doctoral students and faculty authorities, but Holmes began amassing the cache of guns and ammunition used to carry out one of the worst mass shootings in U.S. history.

The subject was listed on the syllabus for a class called "Biological Basis of Psychiatric and Neurological Disorders."

However, Holmes was by then already slipping out of the rarefied world of intellect and scientific discovery that for much of his 24 years had seemed to be a gift embedded in his DNA.

A look over his Wikipedia page details his demise and many of the symptoms he reported, some would argue, are identical and commonly reported to be part of monstrous mind-controlled human experimentation which history documents include audio and visual synthetic hallucinations.

With Sirhan, Sirhan, Dr. Eduard Simson-Kallas was a psychologist who interviewed the convicted assassin Sirhan Sirhan twenty times after Sirhan was sentenced to San Quentin's death row in 1969. He told the San Francisco Examiner that Sirhan was hypnotically programmed. Sirhan could not be trusted to kill Robert Kennedy because he had always been a loser. "He failed at Pasadena City College. He played the horses and lost. He wanted to become a jockey, and fell off a horse." But, according to Simson-Kallas, he was the perfect choice as a programmed scapegoat. "I see him as an excellent follower willing to risk his life for an idea, not afraid of death. He is a very moral person."

It was Edward Hunter, who best described behavior modification, mind control, and the subliminal influence goal stating,

"The war against men's minds has for its primary objective the creation of what is euphemistically called this "new Soviet man." The intent is to change a mind radically so that its owner becomes a living puppet - a human-robot - without the atrocity being visible from the outside.

An excellent YouTube Video entitled "The Real Manchurian Candidate" by e2 Films shows Sirhan, Sirhan being interviewed.

Many others have professed being under mind control influence such as the Fort Lauderdale suspect who claimed the CIA was controlling his mind months before the shootings and, Aaron Alexis, the Navy Yard Shooter, a DOD Contractor reported he was being targeted by the Navy whom he worked for. He carved "My ELF" on his weapon of destruction, meaning my Extremely Low-Frequency Weapon.

In my case, it is USAF and Navy personnel set up around me, confirmed as involved in this program and I have seen both in uniform and felt the beamed ray coming from the specific locations they set up in and are using.

The military and agencies like the Pentagon's DARPA, combined with intel agencies, progress high-tech experimentation and are designated for the personnel who enforce the global paradigm. Today this is happening in many areas, including creating strategically unified federal, state, and local authorities, shared intelligence institutions, and space-based systems connected to advanced electromagnetic technologies via operation centers and ongoing, methods for total and a complete mind-controlled society.

Those who become threats to the agenda, using our young men and women of the Armed Forces, indoctrinated and trained on advanced military technologies in continued research, are marked for high-tech death. Can you blame this for not wanting the truth known?

Because of programming, with absolutely no one exempt, any and every one can become devoid of human compassion, the highest form of intelligence, becoming brain-entrained, brain-dead puppets useful and perfected for ruthless enslavement for population control dominance and usefulness for ongoing Globalist inspired wars.

The bloodshed and death on the battlefield some likened to blood sacrifices depicted in biblical stories related to serving "Evil in high places."

***

In "Brave New World" an excerpt reads:

"This concern with the basic condition of freedom — the absence of physical constraint — is unquestionably necessary, but is not all that is necessary.

A man can be out of prison, and yet not free — to be under no physical constraint and yet to be a psychological captive, compelled to think, feel, and act as the representatives of the national State, or of some private interest within the nation, want him to think, feel and act.

There will never be such a thing as a writ of Habeas mentum; for no sheriff or jailer can bring an illegally imprisoned mind into court, and no person whose mind had been made captive by the methods outlined earlier would be able to complain of his captivity.

The nature of psychological compulsion is such that those who act under constraint remain under the impression that they are acting on their initiative. The victim of mind manipulation does not know that he is a victim. To him, the walls of his prison are invisible, and he believes himself to be free. That he is not free is apparent only to other people. His servitude is strictly aimed."

— **Aldous Huxley, Brave New World Revisited**

CHAPTER 5

LAPD "Rampart Scandal" Act II

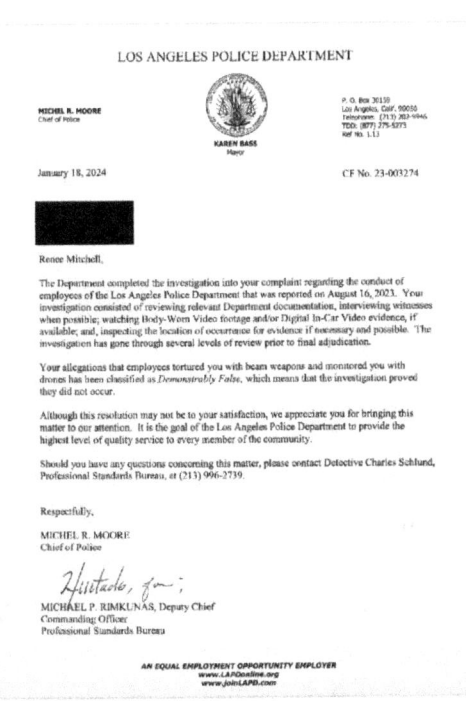

As I came down the homestretch around 4:00 a.m. up for days for publication of this manuscript, the beamed harassment originating from several of the major official setups around me, shown later, became murderously angry using the Active Denial System "Pain Ray". Most of all, the LAPD Black cops were now cowardly concerned about exposure, watching my efforts, with me yet again reporting their horrific low-level activities. As a result, all involved then redirected the beamed focus to my heart as a threat of a beamed heart attack and ongoing focus on my breast. I don't mind telling you that as a perfectionist, getting these books published, my efforts were plagued by many hurtles.

As soon as they saw me leaving to run errands, knowing where I was heading, using thought monitoring technology, one came out of the house across the street where they set up and raced away. As he passed, he was heard saying, "You're dead."

I later learned that he had gone into the poorer side of town and recruited a woman who looked homeless. There is a discount store I like in the area. As I was getting into my car the woman parked near me and glared at me. Her son jumped out of the car wearing a hoodie running towards me towards me in what the pathetic cops thought would be an act of intimidation and insinuation of a gang hit.

On another occasion, around 1:30 a.m., again, while watching me working on this book and the typical torture and beamed death threats, using the technology from the corner house behind, my phone rang. Knowing it was them, and another type of clowning, I did not answer and the call went to voicemail. I did not want to listen to the voicemail as curious as I was and was not surprised when I Googled the phone number and it was from the California State Prison.

Yet again, they had returned to their comfort zone, where their so-called homegirls and homeboys' dwell to recruit someone. These cops are not the educated, sharp Black men who are respectable and honorable. They are those who the

LAPD knew when hiring would be desperate and would kill if they could get away with it or if leadership approved and employment threatened.

As of 2020, the Los Angeles Times, in an article entitled "Hundreds of cases involving LAPD officers accused of corruption now under review" detailed are the horrors still commonly committed by one of the most corrupt police departments in the nation.

Reference Link

https://www.latimes.com/california/story/2020-07-28/lacey-flags-hundreds-of-cases-linked-to-charged-lapd-officers-for-possible-review

One thing is certain, there is no inmate up at 1:30 a.m. and able to use the phone to place a call to my home. It probably was someone working at the jail who they knew.

They briefly succeeded in putting the eBook version of this book on hold with Amazon, making it necessary for me to look at other eBook publishing companies many authors are familiar with. They could not tamper with the printed version because it had the publisher's ISBN attached, however could and did repeatedly hack and sabotage.

Perhaps the most HORRIFIC threat of all as this book materialized was a threat to make one of my daughters commit suicide.

I have heard horror stories of this within the Targeted community, and admit did not believe it initially. Why harm family member I thought. Apparently, many deaths were directed at the target and the targeted person's family members. This is designed to create deep emotional fear and what essentially could be deemed as a form of trauma-based mind control

Destroying a person, family, and any life can be effortlessly accomplished using the same failed tactics they thought would be effective with me. They first destroy all of your hopes and dreams, marriages, support system etc., working behind the scenes. Using subliminal influence documented space-based systems on everyone in any environment, no matter where they are in our country is effortlessly achieved.

In this daughter's case, they destroyed what would have been a prosperous music career as a gifted writer of lyrics and also a beautiful solo artist with great potential which they and many could see. The other two daughters are successful with one a Social Worker and the another of three a Marking Executive. I finally reached a point when they threatened them hoping to affect me to let them live their lives without the high-tech baggage that followed me and transferred to them whenever I was around. This massive program follows you to your death bed.

The Black cops had shown up around her and one romanced her while working undercover for the sole purpose of using her to have me committed, documented in Book IV. The obvious game plan was to blackmail her alleging human trafficking when it was these despicable buffoons running their typical ghetto street games. Family members, which the FBI leadership also know are extremely useful and at times vital for confirmation when having a targets committed to psych wards and highly effective schemes resulting in blackmail is the goal of police set ups.

My children were raised in a conservative Seventh-Day Adventist home, their father a cop and not in Los Angeles, and are pleasantly naïve, beautiful, kind women removed from ghetto mentalities. This is until these hood rat cops show up. As I have documented many times, even today I can't keep these slugs from using the beam focused sexual stimulation patent from between my legs.

After the destruction of a Target's career and life goals, by using frequency manipulation which they also tested on me, can create intense depression where anyone becomes pliable and influenced to sabotage themselves.

It was me they wanted, and the fact is, had proven would use, abuse, and destroy anyone they could, especially if good-looking a perk to keep the truth of this program hidden. IF reportedly, this program is factually experimenting with using mass shooters, and the patented DOD "Voice of God" technology, they hideously lack the empathy to care about anyone or anything.

An old article written about drones with psychotronic technology in the Caribbean reports that this technology can be used to influence suicide. It is accomplished by using beamed mind invasive weapon-equipped on military official drones.

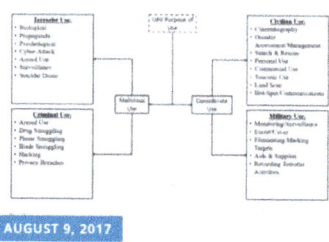

AUGUST 9, 2017

Drone Invasion – 30,000 Drones Approved for US Skies, Drone High-Tech Subliminal Psychotronic Weapons with Mind Invasive Technology, and Beamed USAF ADS "Pain Ray" Torture Subjugation

Reportedly, these drones can be and are responsible for effective suicide experimentation. The details can be read in the titled image of the blog entitled "Drone Invasion - 30,000…" in my Mind Control Technology, You Are Not My Big Brother.blog series.

This program lingers around targets, as stated previously, with human experimentation from generation to generation. This is also a pervasive tactic in the official cover-up creating perceptions of synthetic hereditary illnesses, both psychological or medical.

The drones seen in this blog are the same military type of that follows me around town today and is positioned over my house at this moment and, one of three. The positioning includes a smaller one, weaponized for torture assaults.

The Long-Range Acoustic Device (LRAD) is one example of an official, beamed communication device. It was invented as an off-label weapon which today is used as a sonic pain gun that can project a focused beam of sound at someone's head. Because the Long-Range Acoustic Device (LRAD), can deliver any sound at a distance, soldiers are using it to beam words directly into the ears of civilians determined to be today's enemies, in this case, anyone exposing the massive human experimentation program which can use inaudible subliminal influence and this capability drone equipped.

The technology to those unaware of the technological capability is a nightmare and is designed to make people scared and believe and in fear act out. For example, religious fanatic people can be horrifically duped. As revealed earlier, various types of patented technologies, including LRAD, can put the "word of God" into heads such as the voice of the devil and it can mimic anyone, and in many languages. If God, in the form of a voice inside your head that only you can hear, tells you to kill, what are you going to do? In many, many cases nothing. Some poor souls, however, have acted out instructions as shown earlier after being given deadly directions.

Over the years, since my first book, "Remote Brain Targeting" I was happily a thorn in the side of this program when they left me with no choice, day to day, but to expose what is happening to me, doing so on several social media platforms. At this point, I firmly believe this is the key to my being alive.

The last thing this program wants is confirmation after the fact.

For example, if I tell people on social media, with a large following, that this program is pain beam cooking my head, heart, breast, etc., and I end up with a brain tumor, breast cancer, or stroke, with a clean bill of health, they are the obvious demented culprits and it documented in advance. All of these symptoms are being horrifically orchestrated, as the beamed irradiation of the right side of my right breast progressed in March of 2024 for metastatic breast cancer.

About six months prior, I received a call from a heavily targeted cousin on my mother's side of the family announcing I needed to get checked after she finished chemo. Her mother is my mother's half-sister although I do not have the BrCA gene. Without a doubt she was and still is heavily targeted and behaves as if she has literally lost her mind after being put through the mind control ringer after many years and vicious high-tech abuse. The game plan, believe it or not, is to have a family member phone while everyone is effortlessly monitored today and any location on Earth.

Note that while you sleep and perceived a threat, the 24/7 operation is up and checking every aspect of your life and searching for anyone connected to you, to use and influence and this even includes distant relatives or old friends. With her, her breast cancer battle became a useful opportunity.

Shortly after her phone call, again having not spoken to her in years, the breast cancer beam began from across the street originating from the homes occupied by federal agents, on a different night originating from the USAF home, where I have seen Navy personnel enter, as well as relentless beamed assaults by the Black cops during their shift from early evening until dawn.

I have consistently taken a mammogram for many years yearly. As I approached the age considered extremely low risk when women no longer need the testing the focused beamed assaults escalated. In cases like this, it is not out of the question. Before she barely hung up the phone, as this operation listened, the beamed breast cancer irradiation began. Strategically, these hideous individuals hold the obvious belief, in this case, that if I and when I report that they slow cooked the tissue under my arm and side breast, they can counter this, "She was crazy, breast cancer is hereditary in her family." I have seen many federal agent inspired tactics like this over the years and each effort backed by the promotion of the mental illness tag.

These hideous human monsters, again and again are 100% using beamed energy weapons to create deadly illnesses and the breast focused irradiation left burning pain with their focus on both breasts. I found myself thinking, what would happen if I wore deodorant which typically has aluminum in it? If so, they likely would have quickly achieved their goal by beam cooking this toxic chemical under my arm and the side of my breast.

In one of the blogs on my YouAreNotMyBigBrother.Blog website, links, detailed later entitled "Slow Kill of Undesirables..." Katherine Albrecht, a Harvard Graduate reportedly was given not only a brain tumor but allegedly breast cancer as a whistleblower reporting courageously Radio Frequency Identification Chips (RFID) and covert DNA mass surveillance efforts.

There is a 2015 YouTube CBS video interview, entitled, "Dr. Katherine Albrecht reports 'DNA Tracks'" where she details the diabolical use of bioweapons that can be sprayed on food which would then allow monitoring, tracking and surveillance of everyone in reported in 2015. It appears her suffering also escalated in 2015.

Frankly, the people involved in this program are no better than people locked up in the prison system and even worse because their psychosis has evolved due to government approval.

Another tactic with me, is the hope to exploit side of effects of medication a person takes. For example, Thalidone, which is a Diuretic. In my case, once they learned of the serious side effects, although it has been taken with no issues for years, strategically they decided to create beamed effects. In my case, a report of Chlorthalidone when combined with another drug caused that caused blindness which believe it or not resulted in focus on my eyes.

Again, these people, running the egregious program do not set up around people exposing their outright high-tech cruel official crimes to sit watch you.

They are on an official mission!

It is a documented fact that there are far more women targeted in this program than men which speaks volumes to the mind control motives of psychotic deviants. For obvious they enjoy watching women far more than men. This program is not set up around many targeted today to watch being exposed. They are around to close the troublemakers down by any means necessary. While you sleep, the unified agencies running this program are up strategizing, comparing notes, and searching for strategic methods of silencing where the finger will not be pointed at them after the target's demise.

There were also frustrating attempts to sabotage, erase portions of, and delete this manuscript five times, and insert typographical errors. However, I persevered by transferring it to an external hard drive. I learned from the exact tactics while publishing the six other books in this series, and was prepared for the inevitable.

Of course, the Handlers of the low-level cops and basic military personnel want to know exactly what I wrote. To accomplish this, they hacked my computer and saved the manuscript on their external hard drive for leadership Fusion Center viewing. As a result, several times after I closed the manuscript down for the night, and woke the next day hoping to pick up where I left off, there was nothing to open. Instead, the below pop-up appeared.

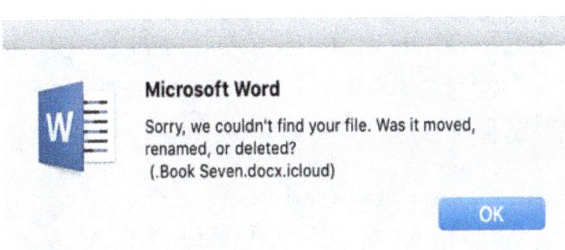

In the eBook version, the hyperlink to this chapter detailing identical corruption by Black Cops is reminiscent of LAPD's infamous "Rampart Division Scandal". So desperate to stop publication of this book, while outright attempting murder, while in production was trashed many times, after hacked. The hyperlink did not click to this specific Chapter that is a mandatory for publishing. This means it could not be uploaded for distribution although the other nine chapters automatically hyperlinked flawlessly, as all had been set to do.

As electromagnetic technology continues to grow and its use, high-tech capabilities of the electromagnetic spectrum have gotten even more sophisticated in this case, aimed at the destruction of what I was writing. They were keeping step with each keystroke and, again, were rearranging words, inserting grammatical errors, and deleting the images of the very real official surveillance setup in my community.

Electromagnetic energy can be used for many things, including tampering with computers and iPhones in real-time. A Business Insider India article dated, July 26, 2021, named, "The insane ways your phone and computer can be hacked and even if they're not connected to the internet" states:

"Both the US and the USSR have spent decades looking into the electromagnetic radiation that an electronic device emits. Kaspersky Lab writes that once a device is plugged into a power line it "generates electromagnetic radiation that can be intercepted by proven technologies."

"Now people have figured out how to harness this information to track keystrokes. Writes Kaspersky Lab

Keystrokes can be remotely tracked with high accuracy at the 67-feet (20-meter) distance by using a homemade device that analyzes the radio spectrum and costs around $5,000. It is interesting to note that the attack is equally effective against common cheap USB keyboards, expensive wireless keyboards with a signal encryption, and built-in notebook keyboards."

Detecting electromagnetic radiation

I, for one, am a witness that this technology is used by official personnel set up in houses on both sides, front and back, who can insert and delete, and rearrange information.

When I work on creating books, the best time for me is during the wee hours of the morning while in the Alpha frequency zone. Alpha waves, with an assist from theta and gamma, are perfect for writing. To get the best work done, you need to be relaxed, with the anxieties dialed down, no telemarketer calls, and your problem-solving and analytical brain turned down too low while everyone sleeps. Because of the monitoring and ongoing attempts to sabotage, I also have to work as fast as possible.

"She's gone!" consistently rang out through the walls from the setups, using another type of beamed communication technology again from one of three drones positioned over my house 24/7. They were upset watching me recover the document they thought they had permanently deleted, which I saved to several locations just in case.

An example of the through-the-wall system used for harassment is detailed below in the Ted.com video link.

This system was developed by Woody Norris and again is one example of various types of technology used for beamed harassment through the walls from neighboring locations inside the target's home.

Hypersonic Sound And Other Inventions

Ted Talk Video Link

https://www.ted.com/talks/woody_norris_hypersonic_sound_and_other_inventions

"She's gone" has repeated over and over again, for years as usually as both the cops and military personnel watched me working.

"Really?"

"This is foolish," I thought as the through-the-wall system beamed the harassment inside my home.

I say this because I made it a point to alert every official in Los Angeles County to what was happening around me, by filing a complaint with the Los Angeles Police Department, Chief of Police. As if they did not know, I detailed the weaponized torture, threats, and stalking by these cops. I also sent complaints to the DOD Inspector General of the Air Force, Navy, and Secretary of Defense.

Frankly, I was fed up.

When I filed the complaint, I also sent images of the locations involved in the surveillance and targeting operation around me with the community prospering revealed by allowing use of their homes. I have seen neighbors get home improvements, new cars, and in one of the homes, where several military personnel used as their training Hub, with the Lockheed Martin trainer, mortgage payments to the homeowner after them move in and I saw the family move out after leasing his home to this program and right after I first saw the FBI enter.

The problem they have with me is that I decided a long time ago that my top priority and number one duty, above all others, is to be constantly aware of my surroundings and observant at all times. As a result, I witnessed familiar faces moving into neighboring locations who are, to this day, the same people who thought I would not recognize them. They were the same cops, the same military personnel who followed me here from where I moved, and some of the same Federal Agents 18 years ago now pretending to be the homeowner or family members of the homeowner visiting.

Not only had I sent a complaint to the LAPD Chief of Police but also the District Attorney, the Attorney General of California, Los Angeles City Council, as the death threats and the harassment continued. There were also beamed announcements of their purported engineering of my assassination. This is easily accomplished using others as mind-controlled puppets, and part of the ongoing daily fear campaign, although key leadership in Los Angeles was now aware.

I observed the Black cops out of their jurisdiction being trained on military weapon systems and devices and all trained by one major DOD contractor, in this case, the Lockheed Martin training team. Without a doubt, Lockheed Martin is heavily involved in field training of electromagnetic weapons across the USA and globally documented later.

The response to my complaint is the letter image that begins this Chapter and it is a false fabrication from a person employed by the LAPD who wrote it and signed "for" the Chief who sounds downright crazy.

First of all, I live 60 miles outside of Los Angeles in the jurisdiction of the Los Angeles County Sheriff. If anything at all was done, in the investigation he reports as the action taken, it would have been conducted by the Sheriff's Department which did not happen. You don't see LAPD patrol cars driving around in a city, again 60 miles away. They are instead hiding in the locations they are using and working in the field.

Although these cops are officially set up, secretively in the Sheriff's jurisdiction it is by approval of the Los Angeles County Fusion Centers for this tortuous mission. Their undercover efforts surpass the criminal activities of the African American cops jailed because of the Rampart Division Scandal involving their association with the "Gangster Rapper" Suge Knight, destruction of evidence, drugs, and murder.

This is the reason for the coverup, at all costs, and the pacifying letter denying essentially denying they are set up and the person signing the response letter detailing outright lies saying my complaint was investigated.

The last thing LAPD would do is let this information get out, much less that a police department with a history of egregious corruption is now using mind invasive, mind control, behavioral modification and psychophysical, military-grade weaponized drone technology for torture, and hiding in a field operation in another city. One night I woke with one of these cops watching in real-time as I slept, around 3:00 a.m., from across the street, trying to block the pain ray with protective material and him saying, "That's not gonna help you" then intensified burning of my skin by the dielectric heating of this weapon.

Below, is a link to a YouTube Video captured by my home security camera of the setup of the home across the street the cops use, with two behind. The video includes a view of the patented flat screen certain neighbors are provided in similar images. Typically, you can hear the technology scratching around in the wall from specific directions before the powerful beamed assault. This technology is being used repetitively on me as this operation rotates shifts for sadistic pain.

Renee Pittman's YouTube Video

The Official Murder for Profit Club Set up Around Human Rights Advocate and Author Renee Pittman

https://youtu.be/UUTbN8tqEsE

The focus on my right leg is their effort at permanent crippling because of the awareness of two surgeries to the right ankle and hip with both directly related to this targeting program, and anger at ineffectiveness and now this seventh book.

Demonstrably False" the LAPD person wrote in his response letter. This means that what I reported as the use of this type of patented technology and torture documented to be in full covert use today, and the accurate use of these weapons, across the USA and globally, and the LAPD belief that it cannot be proven and this to their advantage.

These unified operations around many across our nation are for high-tech experimentation designed as investigations and overseen at a level far above the empty heads of these lowly cops.

The Joint Resource Intelligence Center (JRIC) covers six counties; Los Angeles, Riverside, San Bernardino, Santa Barbara, San Luis Obispo, and Ventura, including nearly 40,000 square miles of territory, 16 million residents, and nearly 200 public safety agencies, including law enforcement, the fire service, and public health agencies. They know exactly what these cops are doing, as the cops repeated "THEY want her" meaning the leadership.

In an article by Brennan Center for Justice Article entitled, "How Government Fusion Centers Violate Americans' Rights — and How to Stop It" dated December 15, 2022 is detailed in the except below.

The Biden administration and Congress must impose oversight and accountability on the state-run centers.

A federal jury awarded $300,000 this month to a Maine State Police trooper who was demoted after blowing the whistle on privacy violations at the state's intelligence fusion center. The federal government spurred the development of fusion centers after 9/11 as a means for sharing counterterrorism intelligence among state and local governments, as well as select private entities. The facts revealed during this trial add to a mountain of evidence that fusion centers require greater regulation and oversight.

The trooper alleged that the Maine Intelligence and Analysis Center, 1 in a network of 80 fusion centers operating across the country, was illegally collecting and sharing information about Maine residents who weren't suspected of criminal activity. They included gun purchasers, people protesting the construction of a new power transmission line, the employees of a peacebuilding summer camp for teenagers, and even people who traveled to New York City frequently. The whistleblower also claimed that fusion center supervisors pressured him to illegally share sensitive FBI information he had access to because of his position on the Joint Terrorism Task Force.

Though fusion centers are operated by state and local governments, the Department of Homeland Security and the Department of Justice provide funding, access to federal intelligence systems, and assigned personnel from the FBI and DHS.

The fact is numerous experiments which are performed on human test subjects in the United States are considered unethical because they are performed without the knowledge or informed consent of the test subjects. Such tests have been performed throughout American history, but some of them are ongoing, which many confirm today.

The experiments include the exposure of humans to many chemical and biological weapons (including infections with deadly or debilitating diseases), human radiation experiments, injections of toxic and radioactive chemicals, surgical experiments, interrogation, and torture experiments, tests that involve mind-altering substances, and a wide variety of other experiments and today little known mind control and behavioral modification patented technologies. Many of these tests are performed on children, the sick, and mentally disabled individuals, often under the guise of "medical treatment". In many of the studies, a large portion of the subjects were poor, racial minorities, or prisoners or kids in the Foster Care system.

1. Most would not be surprised by the creative methods official corruption will try to silence someone exposing them, when they believe their livelihood and employment risks are at stake. They will seek to destroy anyone involved in tarnishing their false public persona, which they value as the shield for corruption and in some cases this case the hideous veil used to hide evil.

In mid-January of 2024, outrageously, there was a masterful, expertly crafted diabolical plan, and admittedly one of the best to date. The Psyop tactic was designed to convince me that I have cancer.

After having a positive fecal test, on January 2, 2024, I requested another from my influenced physician, which was proven by her manner, several times during appointments, then after several incidents becoming blatantly obvious. I took the test again on January 18, 2024.

I did this because when I did the first home test, I knew that I had been constipated the day before and saw a drop of blood on the tissue with no obvious signs of blood ever in my poo. I had been doing Alternate Fasting for months and had not been drinking enough fluids and straining desperately each time during the bowel movement to the point of shaking to expel.

There was anal fissure damage resulting from the straining to push out rock-hard creations done many times, and afterward a feeling of excruciating pain right before exiting at the opening of the anus. This was before I discovered the wonders of MiraLax.

It hurt and the hard poo felt like glass cutting when exiting. I did not doubt that this is where the drop came from.

Bear in mind there is nothing that a person does or thinks that is not watched in real-time under this specific type of high-tech surveillance, and thought deciphering is a must.

As I sealed the mail-in FIT test, I had this on my mind while the "Thought Police" listened. Again, and again, this program would be fruitless without the ability to read subvocal thought. Otherwise, they could only watch in real time and torture you. They need to hear what you think for many reasons, with their effectiveness being one.

The fact is the cops, agents, and military personnel were self-serving and also were desperate to please their leadership, who were demanding my silence. They listened, believing they finally had the perfect plan, hoping to convince me of colon cancer, and also recruited hospital staff to play along.

Do not think for one moment they would not send me through deadly chemo to make me sick by injecting unnecessary poison into my body to fix an issue that did not exist.

Under 24/7 surveillance for years, they were there when I registered that I had seen a drop of blood on the tissue one day before sending in the test and that I thought that this test might be positive and I felt should not have been sent in. I played right into their sinister hands. If it was not positive, they were going to make sure I thought it was for what panned out to be a cancer scheme.

There has not been one time over the years when I went to the VA hospital for doctor appointments that they had not shown up, telling everyone they were investigating me and turning people against me. I have always been gregarious and talked with everyone all my life. To their dismay, people liked me. This was not good for the narrative they were promoting.

What I did not realize is how horrifically and deeply this program was searching for various types of methods to silence me, again strategically and permanently.

One of their ongoing goals has always been to take my book series, blogs, and website down. As they watched me working on this book, I had yet again become a decisive threat, this time with an eye-catching, accurate title of their real motives and images of this program set up around me.

I have taken this test yearly and never had I gotten results within two hours from the backed-up VA laboratory. or it followed up by an email from one of the VA medical staff whom, again, I realized in early 2023 were playing along with the Los Angeles Police Department corrupt cops who came out during each appointment and lurked in the background.

After an email from this nurse, looked altered, I visited the ER that night. Troubling was the unusually quick test result.

As it turned out, the second result which she reported in the email as positive was not. I checked while at the hospital that night in the Hematology Clinic. When the lab technician pulled up the results, the system listed the test as still processing. Yet I was told that it was processed and positive.

Can you imagine if I believe this? Apparently, the human monsters running this program did?

I also had a CT Scan of my abdomen at the ER, which showed no issues except a small amount of compacted poo which the MiraLax released. Yet these men set up around me as soon as I made it home late that night began a fierce beam zapping to my abdomen. It was obvious they were trying to create this deadly condition which they hoped would finally achieve their goal of shutting me down and what they hoped to energy weapon beam create confirmed by the false positive lab result.

The attacks on my abdomen were so intense that night that one powerful zap woke me up.

An abdominal CT scan can diagnose various medical conditions such as stomach inflammation, GERD, impacted bowel, gastritis, polyps, intestinal problems, diseases of the small bowel, diseases of the colon, cancer of the renal pelvis, cancer of the ureter, colon cancer, lymphoma, melanoma, ovarian cancer, pancreatic.

Mine revealed nothing except again, a small amount of poo stuck and it was likely residue from the extreme constipation.

The CT Scan also showed no **Diverticulitis**, also called **colonic diverticulitis**, and is one of the gastrointestinal diseases characterized by inflammation of abnormal pouches—diverticula—that can develop in the wall of the large intestine. Symptoms typically include lower abdominal pain of sudden onset, but the onset may also occur over a few days. There may also be nausea, diarrhea, or constipation. Fever or blood in the stool suggests a complication. People may experience a single attack, or they can experience repeated attacks, or ongoing "smoldering", diverticulitis.

While the Black cops told hospital staff they were investigating me, they had minimal influence. The federal agents were the ones with the real influence and power of persuasion. In this case, getting people to agree to be used and get involved at this level of deception.

I have witnessed over the years, diabolical plans with them operating in the background with judges and other influential people who have crossed my path during the duration of the official leadership around me overall.

With military subliminal beamed influence and backup, many have agreed to play along. They also told people I am a drug dealer or addict, etc., and later discussed, to mobilize the community, which is a good one. No one wants drugs in our community of 5% Black people and neither do I!

This narrative is used to motivate Community Volunteers stalking, who think as I run errands am distributing drugs, to the point that I put my website on my license plate on my car and a picture of perfect teeth on the website when opened which reveals a person not fitting the narrative they are selling to mobilize community help. They are telling the community one thing. However, when they Google my website, a completely different story emerges, revealing identical tactics for everyone, from all walks of life, and massive inhuman suffering and strategic murder nationwide, and the fact that they are covering up hideous high-tech human experimentation.

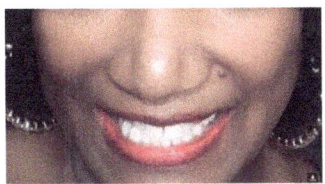

I am in perfect health, no high blood pressure, cholesterol or diabetes.

I also ordered a home FIT test kit and did it one morning after waking, I was not surprised that it showed negative results for zero blood in my stool.

The question is, it the test reliable?

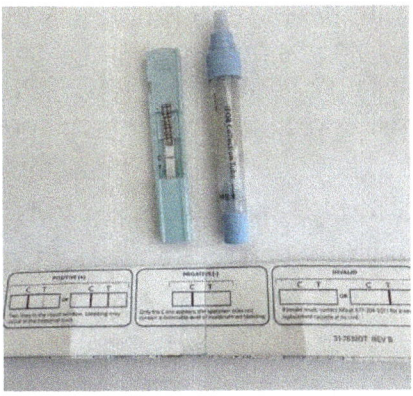

According to the Cedars-Sinai Blog it is and similarly accurate as home pregnancy tests.

Before I left the hospital, that night, the ER doctor told that after reviewing the CT Scan it was likely another cause if anything at all. I have for years had a benign fibroid common to at least 50% of all women he believed was the actual cause any discomfort and nothing alarming to worry about.

As I drove home, and it proven that I was lied to about the second FIT test and it confirmed that the second test was still pending, I realized again, and understood the horrors those involved in this program were prepared to take and take me through, what they also considered as a military Psychological Operations for my strategic demise and ultimate my death.

Our military has been unleashed for human experimentation on the civilian population for DECADES, with many official agencies aware of this, connected and involved with ongoing mass social and population subjugation control at the helm of inhumane technology one bothersome person like me and others are 100% expendable.

Although there are many honorable people in life, and in the military, cops, and federal agents, if looking for heroes, look elsewhere! They are not the ones involved in human experimentation today, and heinously today experimenting on American citizens of all races across our country.

"Thank you for serving!" the global technocratic, mass population control paradigm while under official orders and lives destroyed! Military energy weapon assaults using the USA Civilian population as lab rats is 100% factual and accurate.

As a precaution I decided to have the Colonoscopy and as usual I had to inform the doctor of the details surrounding the false positive FIT test. While doing so, the military drone that followed me there scratched around in the ceiling to focus and watched and listened in real-time.

I could hear the Black cops commenting, "She's going to tell him" using the beamed system. After I told him, they watched me give him my business card with my website, blog and book information. I told him, as I parted with the Colonoscopy upcoming I have a forthcoming book, entitled: Targeted: "If I Die, This Program Killed Me!" I did not mention they have been beam zapping my abdomen trying to create issues.

As I drove off, again using the, PATENTED drone beamed communication system, I heard these monstrous hideous men say, *Well, that's not gonna work!* I thought, *I guess it's back to the drawing board for them and Act III.*

Time will tell what's next in their determination to shut me up so they can continue doing to many what they attempt with me heinously.

CHAPTER 6

Organized Stalking is No Delusion

DARVO stands for "Deny, Attack & Reverse Victim for Offender."

The perpetrator or offender will Deny the behavior, Attacking the individual doing the confronting and Reversing the roles of Victim to Offender. They will use this to restrict your thoughts and behavior by blocking and interfering with your memory and thought process with their impulses fabricated injections of thoughts and memories. All of this is done with highly classified technology created by the government. You must not fall into their ridicule factors or you'll become the monsters that they are. Our whole lives are destroyed because we somehow became a threat to their Orwellian rogue evil narcissistic agenda.

Domestic violence incidents — including murders — often begin with abusers stalking their victims, including officially. Many experts report this program as Hive Mind military experimentation, group programming, and again, they can be programmed to kill!

A clue to who is running the reported massive stalking program nationwide is revealed by once insider ex-FBI supervisor of the Los Angeles FBI Field Office, Ted Gunderson, deceased who became a "Targeted" whistleblower after he left the Bureau shown in an authentic excerpt of a letter he wrote which is the second image below.

Symptoms of Microwave Illness
https://www.microwavedvets.com

Headaches	Difficulty Concentrating	Tinnitus
Dizziness	Memory Loss	Hearing Loss
Nausea	Brain Damage	Irregular Sleep Pattern
Skin Rash	Mood Disorder	Insomnia
Itchy Skin	Personality Disorder	Chronic Fatigue
Burning Skin Sensation	Increased Irritability	Deteriorating Vision
Tingling Sensation	Decreasing Trust in People	Pressure in/behind eyes
Tremors	Depression	Eye Damage
Muscle Spasms	Anxiety	Cataracts
Muscle and Joint Pain	ADHD/ADD	Immune Abnormalities
Restless Leg Syndrome	Digestive Issues	Altered Sugar Metabolism
Foot Issues	Abdominal Pain	Asthma Attacks
Low/High Pressure	Enlarged Thyroid	Bronchitis
Facial Flushing	Hair Loss	Pneumonia
Dehydration	Testicular/Ovarian Pain Low	Inflamed Sinuses
Body Metals Redistribution	Sperm Motility	Chest Pain/Pressure
Leukemia	Miscarriage	Heart Arrhythmia
Lymphoma	Electromagnetic Sensitivity	Heart Palpitations

3. I have read the Complaint in the current action of Mr. Keith Labella against F.B.I. and D.O.J. It is my professional opinion, based on information, knowledge and belief that the information sought by Mr. Labella in this F.O.I.A. suit regarding "gang stalking", "gang stalking groups" and "gang stalking methods" reasonably describes an ongoing, active, covert nationwide program that is in effect today, and, based on my investigations and experience, has been operational since at least the early 1980's. Since the 1980's gang stalking has increased in scope, intensity and sophistication by adapting to new communications and surveillance technology. These programs are using the codenames Echelon Program, Carnivore System, and Tempest Systems. The Echelon Program is administered by the N.S.A. out of Fort Meade, Maryland, and monitors all email and phone calls in the world. Carnivore System is administered by the N.S.A. out of Fort Meade, Maryland, and can download any computer system without being traced or otherwise known to the owner. Tempest Systems can decipher what is on any computer screen up to a quarter of a mile away. These programs are negatively impacting thousands of Americans and severely abusing their civil rights on a daily basis.

Law enforcement officials say harassers are increasingly using technology to harass victims — tracking their every move in a mobilized community harassment program by an official app alerting mobilized groups to the target's whereabouts in a nationwide network, which apparently. they are unable to put a stop to.

However, this type of stalking is reported to be mobilized by law enforcement, at all levels, Federal, state, and local, joined by military personnel using Psychological Warfare tactics, and this is why it has not been publicized as reality and shut down.

The fact is the delusional mental illness tag of those reporting this nationwide network, developed from training using intentional, bizarre tactics known as Street Theater with everything watched in real-time.

Across the United States and globally Targeted Citizens have united, revealing this horrific, massive, global f psychological operation with the same Standard Operating opposition to the mass, social, and population control agenda. These tactics are nothing new historically and today and are a version of a modernized Stasi Zersetzung mentioned earlier and defined below.

Zersetzung in German was used for decomposition and disruption) was a mass psychological warfare technique used by the Ministry for State Security (*Stasi*) to repress political opponents in East Germany during the 1970s and 1980s.

Zersetzung served to combat alleged and actual dissidents through covert means and targeting using secret methods of abusive control as well as psychological manipulation to prevent anti-government activities.

People were commonly targeted on a pre-emptive and preventive basis, to limit or stop activities of dissent that they may have gone on to perform, and not based on crimes they had committed.

Sensitizing methods were specifically designed to break down, undermine, and paralyze people behind a facade of social normality in a form of silent repression.

Erich Honecker's succession to Walter Ulbricht, as the First Secretary of the Socialist Unity Party of Germany (SED) in May 1971 saw an evolution of operational procedures (Operative Vorgänge) conducted by Stasi away from the overt terror of the Ulbricht era towards what came to be known as Zersetzung (Anwendung von Maßnahmen der Zersetzung), which was formalized by Directive No. 1/76 on the

Development and Revision of Operational Procedures in January 1976.

The Stasi used operational psychology and its extensive network of between 170,000 and over 500,000 informal collaborators (inoffizielle Mitarbeiter) to launch personalized psychological attacks against targets to damage their mental health and lower chances of a hostile action against the state. Among the collaborators were youths as young as 14 years of age.

Whether it's a boss, a partner, a parent, a friend, or a unified cadre of official government agencies or organizations, the goal is to create self-doubt and a perception that you are losing your mind and this is a highly effective, time-tested, successful tactic used by governments, covering up today mass human experimentation, and is highly effective with thousands coming forth and exposing experimentation combined with strategic, Psyops tactics by design.

Why is this happening today? What is the common thread that unifies innocent citizens growing into millions reporting this official Organized Stalking covert crime?

A look back in history reveals identical tactics can be traced back to antiquity in efforts designed to ensure a controlled global society. In 1 Peter 5:8-9 it reads:

Be sober-minded; be watchful. Your adversary, the devil, prowls around like a roaring lion, seeking someone to devour. Resist him, firm in your faith, knowing that the same kinds of suffering are being experienced by your brotherhood throughout the world.

Why are police and government officials doing nothing?

Reportedly Gang Stalking is a covert government military intel and police program unified. It is similar to and identical to the COINTELPRO of yesterday or Red Squads, and it's being used on a lot of innocent people to ruin them and make them look crazy. This program is fueled by Military Reserve personnel within the U.S. now given domestic action.

The Reserve was created to provide and maintain trained units at home while active-duty service members are deployed. With the exception of the Space Force, each military branch has a Reserve component under its command, which is available for active-duty deployment in times of war or national emergency. Other Reserve branches provide backup for the Space Force as foot soldiers inside the U.S.

As outlined in 10 U.S.C. § 9081 and originally introduced in the United States Space Force Act, the Space Force is organized, trained, and equipped to: Provide freedom of operation for the United States in, from, and to space; Conduct space operations and protect the interests of the United States from space. This is the who is running the U.S. Space Force, not only in the U.S. alone but goes hand in hand with thousands of militarized drones used at home and quietly approved by U.S. Congress.

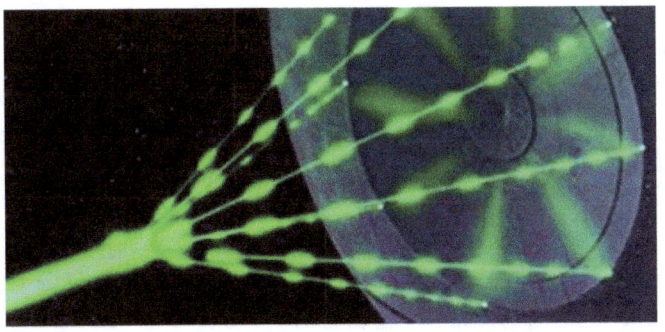

Nearly 190,000 Army Reserve Soldiers and 11,000 Civilians are present in all 50 States, and five U.S. territories, and this is the Army alone.

For example, Jim was placed under covert investigation by his employer because of his whistleblowing. Mistakenly, he termed it Gang Stalking, which made him sound crazy.

It is all about government disinformation using civilians to help with stalking and monitoring innocent people and is powerfully discrediting. Within this massive program are perfected tactics of covert and high-tech harassment designed to make the target look mentally unstable when reported.

These tactics have proven successful within the cover-up through many generations. Today, the unified global community declares the game is over!

Targeted Citizens are awake and prepared to fight an intelligent battle using exposure and are undeterred as the opposition continues in denial.

Exposure remains the strongest weapon because it is the absolute truth and ultimately powerful, especially when combined with sheer determination.

Thousands are standing up and will not be swayed, no matter what is said. You should not care. When this happens, the result is that the truth becomes a major, gradual force for good.

It is important to remain steadfast knowing that if many people are saying the same thing, people will begin to listen, and this includes those recruited.

Hopefully, they too will awaken to the truth that all of this is pure, human experimentation, with no one exempt, including their programmed hive minds.

The scientific excerpt abstract below describes the Artificial Intelligence connection to the dynamic of Hive Mind application and human experimentation. It is derived from the article: *AI* **2022**, *3*(2), 465-492; and link: https://doi.org/10.3390/ai3020027

It was submitted April 5, 2022 revised: May 4, 2022 and accepted: May 5, 2022 then published May 16 2022

(This article in its entirety belongs to the Special Issue Standards and Ethics in AI)

Excerpted Abstract

Insect swarms and migratory birds exhibit something known as a hive mind, collective consciousness, and herd mentality, among others. This has inspired a whole fresh stream of robotics known as swarm intelligence, where small-sized robots perform tasks in coordination.

The paper also showcases a framework for how cybernetic hive minds have come into existence and how the hive mind might develop in the future. It also discusses the implications of these hive minds for the future of free will and how different malfeasant entities have used these technologies to cause problems and inflict harm through various forms of cyber-crimes and predict how these crimes can grow in the future.

Keywords:

[AI](); [hive mind](); [ethics and AI](); [human-centric AI](); [novel interface]()

Yes, Organized Stalking can be more powerful in its effect than simple bullying, but the problem is the same.

These groups will often continue to do what is useful until the target stops acknowledging the tactic as emotionally effective, and more importantly, understands that everyone and everything is monitored in real-time in this new mass surveillance paradigm.

You must accept the motive, and the purpose, and learn to cope by responding positively or by simply ignoring the foolishness designed to entrain your mind and maintain your focus. The fact is, there are far more pleasant things to redirect your mind to.

With me, it took something so simple as my website on my vehicle plate to counter what the community was told about me when sent out to follow me around town in some cases, used as a pathetic death threat. You can't be scared and expose this program at the same time.

When you change how you react to anything in life and, in this case, the modernized Zersetzung group dynamic, its power is gone, and the victory is yours and not theirs. Every response should aim at meeting this goal of eliminating the effects of Organized Stalking through awareness.

Don't give up! The truth is fast becoming undeniable. Be assured in faith and trust that the truth will prevail and inevitably always has.

You, my friend, are not alone!

To survive, you must be aware of how targets are duped with the mental illness tag, with various tactics designed to make anyone exposing this program look crazy as part of the coverup. This is vital and this program's goal. Don't forget this.

The stalking effort is 100% part of Space Force human experimentation and the technological implications behind mind-invasive technology implemented for massive military experimentation with Remote Neural Monitoring. Remote Neural Monitoring satellite systems have 400 channels, each channel capable of remotely reading one human mind per second.

The satellite system can project electromagnetic precision beams at each target (up to 400 individuals at any time) and place each targeted person within an electromagnetic field.

They can then use this EMF to extract EEG data from the human brain. This data is then transmitted from the satellite back down to a receiver and decoded with a computer system to extract sub-vocalizations, visual, auditory, and sensory information to be interpreted and presented before official agencies involved in this type of human experimentation led by military technology, again, Federal Agents, military intel, and local law enforcement officials on a monitoring screen.

Every human being gives off their own bioelectric magnetic resonate frequency, which corresponds with their biometric ID. This is a fingerprint that identifies and distinguishes each individual. Almost like a unique radio frequency that they can tune in to with the goal of monitoring on a massive scale.

Electromagnetic frequencies and extra-low frequencies are sent to the brain from these satellite systems to trigger what is called an evoked potential within the brain of the targeted person, including puppets who are recruited for organized stalking. An electric and magnetic reaction in the human mind creates a spike in brainwave activity and an interference wave pattern about the transmitted EMF and ELF from the satellite system that can be mapped out and deciphered.

Electromagnetic signals capture different frequencies and are simultaneously transmitted to the brain of the subject in which the signals interfere with one another to yield a waveform that is modulated by the subject's brain waves and also allows computer-aided mind reading. The interference waveform, which represents the brain wave activity, is re-transmitted by the brain to a receiver, where it is demodulated and amplified. The demodulated waveform is then displayed for visual viewing and routed to a computer for further processing and analysis.

The demodulated waveform also can produce a compensating signal which is transmitted back to the brain to effect a desired change in electrical activity. To transmit communications back down to the targeted individual, a corresponding EMF signal is beamed directly into the auditory cortex of that individual from the RNM satellite system and today psychotronic equipped official drones, and the person can hear the official or the computer system feeding back communications with them in real-time. They can also apply other methods of responding to such communication.

They can use ground-based, microwave cellular towers with directed microwave technology that uses microwaves with the wavelength of acoustic waves, which essentially transmits data from what is believed to be nonfunctional GWEN towers directly to the cochlea of the individual to produce a different effect.

This again is called LRAD or S-Quad technology and is more commonly used by the U.S. military against innocent civilians in an activity founded on Stasi Zersetzung to make the individual enter psychosis. The technology mimics mental illness.

Scalar beam technology and high-frequency radio transmissions can also be used to produce similar effects. The BNCI 2020 is using this technique in conjunction with Bio API and psychotronic energies are being used to keep people captive and use them as "subjects for human experimentation" with many a focused target and many playing different consensual roles.

This is the way this again, mass, social population control program is designed.

If you report anything you experience, you will be diagnosed with delusions, and also reports of being officially stalked, which is strategically designed as part of the procedure of this program. Artificial Intelligence will keep a target from sleeping to break the target down as well. Extreme sleep deprivation remains a useful tactic as well. While constantly running the brain of everyone 24/7 it does not stop. At the same time, targets are bombarded with insults and threats, and the stalking puppet is bombarded with negativity about the target. This is accomplished by locking onto different brain centers of the target or those used for Hive Mind Programming, stalkers, and hope to keep people in an angry state and motivated anger against the specific target.

The diabolical result is that the target has to hide powerful negative emotions every day, although many have become successful at transmuting the frequency, manipulation, and assaults by vibrating at a higher frequency. Love elevates positive thought and understanding. You must dig deep for forgiveness to heal.

Governments are conducting long-term lifetime experiments that are, are crimes against humanity. These high-tech crimes are being committed, as open literature evidence confirms reality.

The experiments are cruel and demented they need A.I. to run the experiments and collect the data again on a massive scale. All governments today have this technology, with the U.S. Russia, and China leading the pact.

What we are seeing today is the tip of the iceberg of a fast unfolding mass population global agenda and a similar Cold War high-tech race between Superpowers, that includes scientific progression into trans-humanism and much more. Is humanity obsolete in the minds of Globalists?

MSN.COM
Exclusive: Inside the Military's Secret Undercover Army
Thousands of soldiers, civilians and contractors operate under fa...

"The largest undercover force the world has ever known is the one created by the Pentagon over the past decade. Some 60,000 people now belong to this secret army, many working under masked identities and in low profile, all part of a broad program called "signature reduction." The force, more than ten times the size of the clandestine elements of the CIA, that carries out domestic and foreign and foreign assignments.

They are both in military uniforms and under civilian cover, in real life and online, sometimes hiding in private businesses and consultancies, our communities some of them household name companies…"

Reference Link

https://www.newsweek.com/exclusive-inside-militarys-secret-undercover-army-1591881

CHAPTER 7

The Blog 'THEY' Don't Want You to See!

Military Civilian Targeting
Hiding In Plain Sight

Abstract

Over the last several decades, the range and capabilities of easily available technologies have expanded at an astonishing pace. The beeper gave way to the flip phone, which has largely been replaced by the ¿smartphone,¿ a mini-computer that fits in the palm of your hand and is more powerful than the desktop machine of the 1980s. Paper maps are increasingly rare, replaced by built-in Global Positioning

System (GPS) devices or the ubiquitous smartphone. These and other technologies, which are valuable to civilians and law enforcement alike, also enable a granular view of citizens movements and associations in public over long periods of time at a relatively cheap cost.

Where law enforcement is involved, these powerful new technologies also raise questions about how their use can be harmonized with the U.S. Constitution.

These authentic images accurately reveal the setup around Human Rights Advocate, Author, Renee Pittman revealing, a major, official, military COINTELPRO, and corrupt police targeting operation with several drones used by this program for tracking, monitoring and beamed subjugation assaults.

Know of a surety that only the truth brings this type of official silencing setup into a neighborhood happening because of exposure of official corruption and the Target has become a major threat.

This program actually thought they would operate in secrecy in what continues to evolve into a slow-kill operation. However, as I, Renee Pittman stated previously, it is one of my upmost duties to be aware of my environment and observant of my surroundings at all times. My very life depends on it. You also can hear the optical lens for in home viewing positioning itself inside the ceiling and walls before energy weapon attacks revealing the direction of origin.

I witnessed on family move out, and give full reign of their homes for these types of operations where they receive perks such as mortgage payments, home improvements, new cars, etc., from the DOD Black Budget funding which is in excess of billions of dollars availability.

The images also reveal how the strategically placed military Active Denial System "Pain Ray" antennas are placed in vehicles and also parallel rooms to the target's home, as well as the bright circular lights in upstairs and downstairs rooms for the Sonic Weapon assaults. All of the technology, although reported as non-lethal, a misnomer, when used repeatedly on a target becomes gradually deadly, and can create various cancers, Parkinson's, and more and ultimately strategic death.

In fact, an ex-NSA spy Mike Beck documents his belief that beamed microwave weapons were the actual cause of his Parkinson's disease in a 2017 Washington Post article.

One of the first signs came at the keyboard. Mike Beck, a National Security Agency counterintelligence officer, could always bang out 60 words a minute. But in early 2006, Beck struggled to move his fingers at their usual typing speed. He had to hunt and peck.

Soon after, a brain scan showed why: Beck had Parkinson's disease, the second-most-common neurodegenerative disorder in the United States, behind Alzheimer's. He was only 46 — unusually young for Parkinson's. No one in his family had ever had it. Then, in an unsettling coincidence, he learned that an NSA colleague — a man he'd spent a pivotal week in 1996 with in a hostile country — had *also* just been diagnosed with Parkinson's.

After nightly attacks from two houses behind me, and several in front across the street using the Sonic Weapon, which creates a deadly hummm, hummm, hummm, I woke with my hands twitching and jerking uncontrollably. It got so bad that I had it documented in my medical records at the

Veterans VA hospital also documenting it as the result of this particular technology.

Another example of beamed assaults, while I slept at night or napping during the day, are powerful sporadic zaps to other parts of my body. The beamed hits were similar to the intensity to my abdomen, with both, throat, waking me coughing uncontrollably and my body and abdomen, jerking previously mentioned.

In one of my 112 blogs, (Mind Control Technology - You are not my big brother blog, by Renee Pittman) entitled: Slow-Kill of Undesirable by Beamed High-Tech Crippling...several types of deadly, as well as debilitating illness of individuals who are whistleblowers, activist, or those who pissed someone off in high-places, appear to be plagued by various types of deadly illnesses, Parkinson's, throat cancer, brain assaults, (my head hit regularly) that may have been synthetic high-tech weapon manipulated are detailed.

Although all involved in the setup were zapping me during their rotating shifts from neighboring locations setups for this cause, now their neighborhood operation center, I suffered the most during the shift of the LAPD Black cops who repeated, evert day as their shift began, *THEY* want her.

It is a documented fact that many people go to the Black community to hire desperate Black men to do their dirty work. However, this time they are officiated and put in uniform. They are the type that if not doing this would likely be selling drugs, trying to traffic women more to survive and are not the intelligent honorable type with honorable purpose in life frankly, with military personnel and federal agents no better.

As I have said before, *Birds of a feather flock together*

The same with military personnel used. They are those who have access to White Privilege but do not know what to do with it and sign up for military hero worship and as a result, the truth of this horrific human experimentation program setting people up useless to them! Yes, there are heroes but they are not working this hideous covert destruction program of men, women and children as official Order Followers.

Renee Pittman Mind Control Technology

Slow Kill Blog

Reference Link

Strategic SLOW KILL of Undesirables by Beamed High Tech Crippling, Heart Attacks, Brain Aneurysms, Breast & Various Cancers, within a Unified, Military COINTELPRO Paradigm of Official, CORRUPT, Technological Terrorism. Is it possible?

Official personnel use of garages as operation sites as well. The result is 24/7 official PsyOps, beamed assaults, including military Reserve personnel stalking when the target leaves home which includes drone tracking and monitoring around the community. When the Target returns home the drones take stationary positions around the target's home of the operators.

Military weaponized drones also include psychotronic mind invasive technology, beamed communication systems, for this type of surveillance mission, and are using aerial beamed assaults, patented subliminal technology for manipulative influence of any and every one around the target at any given time with many unaware what is happening.

The images were captured when waking during beamed assaults at night and high-tech torture.

Military personnel and all involved are 100% using military grade weapon systems. The capability for high-tech weapon use and surveillance through computerized weaponized, operating systems connects to satellites to drone space derived, electromagnetic technology.

DOD Contractors training everyone also used deceptive alien infused beams as part of the cover up and again, instead of working in official offices, community locations become their field operation and again, the community paid well for compliance.

Many, many lives have been, and are still being, completely destroyed today in decades of ongoing, high-tech experimentation on reported national scale with specific communities hit hard as human guinea pigs.

This link confirms, human experimentation from the past too today driving home the reality that MKULTRA never ended!

GENERAL COUNSEL OF THE DEPARTMENT OF DEFENSE
WASHINGTON, D. C. 20301

September 20, 1977

MEMORANDUM FOR THE SECRETARY OF DEFENSE

SUBJECT: Experimentation Programs Conducted by the Department of Defense That Had CIA Sponsorship or Participation and That Involved the Administration to Human Subjects of Drugs Intended for Mind-control or Behavior-modification Purposes

Chemicals, Warfare, Mind Control & Behavior Modification Programs Ongoing Eternally

***

When I moved here 10 years ago, the first thing I noticed was that the same USAF personnel assigned to me had followed and immediately setup shop in the garage directly next-door South. They now operate from a house, newly purchased by a USAF person across the street.

A short time later, I learned that the son, and son-in-law next door, the homeowner's family are Navy personnel. I thought "Oh boy" talking about going from the frying pan and into the fire.

The implication of this awareness was instant.

Without a doubt this house would be a primary player in the targeting effort around me due to proximity to my bedroom, upstairs and the perfect coverup as a Navy family home. When I saw the son's, SUV parked in his mother's driveway I knew what to expect that night as I suffered beamed attacks all night. It was used before the son was reassigned to home duty by USAF personnel and in one of the images while setup in the garage put his military shoes outside the garage door to air out.

This technology, again when uses relentlessly is no longer nonlethal as defined in the excerpted link below.

Sonic Weapon

Sonic and **ultrasonic weapons (USW)** are weapons of various types that use sound to injure or incapacitate an opponent. Some sonic weapons make a focused beam of sound or of ultrasound; others produce an area field of sound. As of 2023 military and police. **Sonic** and **ultrasonic weapons (USW)** are weapons of various types that use sound To injure or incapacitate an opponent. Some sonic weapons make a focused beam of sound or of ultrasound; others produce an area field of sound.

Extremely high-power sound waves can disrupt or destroy the eardrums of a target and cause severe pain or disorientation. This is usually sufficient to incapacitate a person. Less powerful sound waves can cause humans to experience nausea or discomfort.

The possibility of a device that produces frequency that causes vibration of the eyeballs—and therefore distortion of vision—was suggested by paranormal researcher Vic Tandy[1] in the 1990s while attempting to demystify a haunting" in his laboratory in Coventry. This spook was characterized by a feeling of unease and vague glimpses of a grey apparition. Some detective work implicated a newly-installed extractor fan, found by Tandy, that was generating infrasound of 18.9 Hz, 0.3 Hz, and 9 Hz.

***

USAF, Navy Testing And Research Of Beamed Weapons

Directed energy weapons (DEWs) are defined officially as electromagnetic systems capable of converting chemical or electrical energy to radiated energy and focusing it on a target, resulting in physical damage that degrades, neutralizes, defeats, or destroys an adversarial capability. Navy DEWs include systems that use High Energy Lasers (HEL) that emit photons, and High-Power Microwaves (HPM) that release radiofrequency waves. The U.S. Navy uses DEWs for power projection and integrated defense missions. The ability to focus the radiated energy reliably and repeatedly at range, with precision and controllable effects, while producing measured physical damage, is the measure of DEW system effectiveness. Conversely, capabilities to increase the resilience or survivability of platforms or Sailors from DEW threats are part of the Counter Directed Energy Weapons (CDEW) program.

Navy Mind Control Research Programs

U.S. Navy research on "mind control techniques" cannot be performed on human subjects without the authorization of the Under Secretary of the Navy, according to a new Navy Instruction (pdf).

"The Under Secretary of the Navy (UNSECNAV) is the Approval Authority for research involving ... severe or unusual intrusions, either physical or psychological, on human subjects (such as consciousness-altering drugs or mind-control techniques)."

The nature and scope of any such Navy research could not be immediately discovered.

See "Human Research Protection Program," Secretary of the Navy Instruction 3900.39D, November 6, 2006 [at section 7(a)(2), page 9].

The information and images in this Chapter are derived from the Mind Control Technology You Are Not My Big Brother Blog, by Renee Pittman.

The goals of over 112 blogs are to bring awareness of human experimentation that continues to operate, literally under the radar. This blog detailing the setup on the WordPress site was repeatedly sabotaged and deleted.

Frankly there is no protection of the beamed electromagnetic weapons being used today however, the R-Tech does, from Home Depot effective block their viewing of the target inside homes. The operators of the system need to see the target.

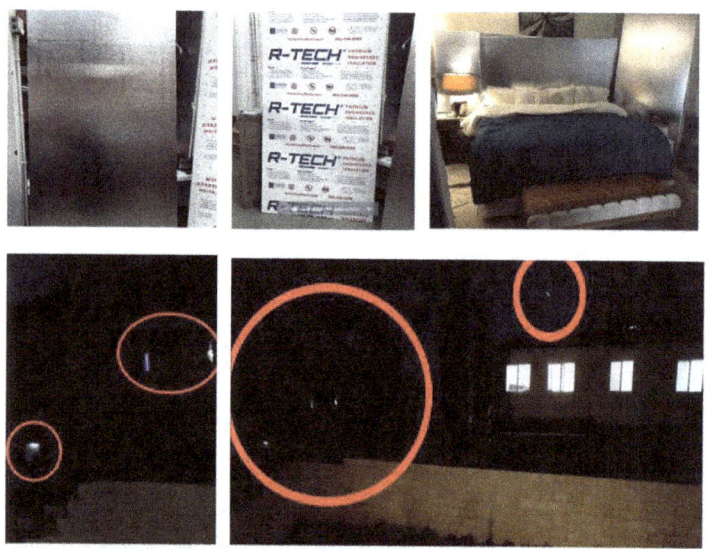

Note that the bottom left image above with two circles, with top right circle revealing the technology that illuminates locations emitting a blue or lavender hue as and also some location illuminated

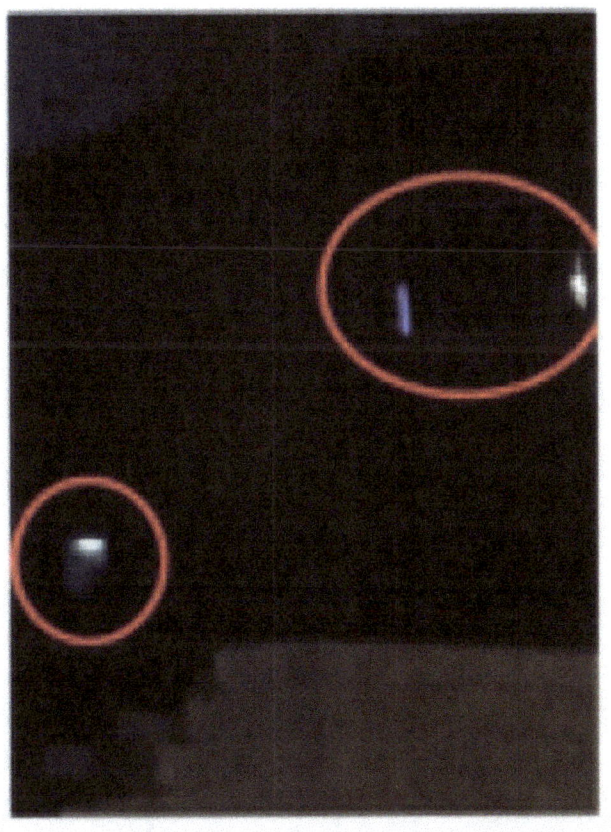

Electronic Concentration Camp

"[A security camera] doesn't respond to complaint, threats, or insults. Instead, it just watches you in a forbidding manner. Today, the surveillance state is so deeply enmeshed in our data devices that we don't even scream back because technology companies have convinced us that we need to be connected to them to be happy."

— Pratap Chatterjee, journalist

What is most striking about the American police state is not the mega-corporations running amok in the halls of Congress, the militarized police crashing through doors and shooting unarmed citizens, or the invasive surveillance regime which has come to dominate every aspect of our lives. No, what has been most disconcerting about the emergence of the American police state is the extent to which the citizenry appears content to passively wait for someone else to solve our nation's many problems. Unless Americans are prepared to engage in militant nonviolent resistance in the spirit of Martin Luther King Jr. and Gandhi, true reform, if any, will be a long time coming.

***

While many are reluctant to expose that our military, who are playing a major role in surveillance and heinous weapon torture and far worse mind invasive technology human experimentation, turned on the civilian population, the fact is the trail leads back to the DOD connection and use of military grade technology and military personnel.

With my life on the line, I owe no allegiance to any organization, agency or uniformed Order Followers today sitting at the helm of advanced technologies in nationwide operations that should be appropriately named the Militarized Police State.

There are at least numerous military personnel, both USAF and Navy, and Reserve, again, as stated, previously assigned to me, at least four cops and several federal agents who think I am a perfect human guinea pigs. I disagree and have proven I for one am not!

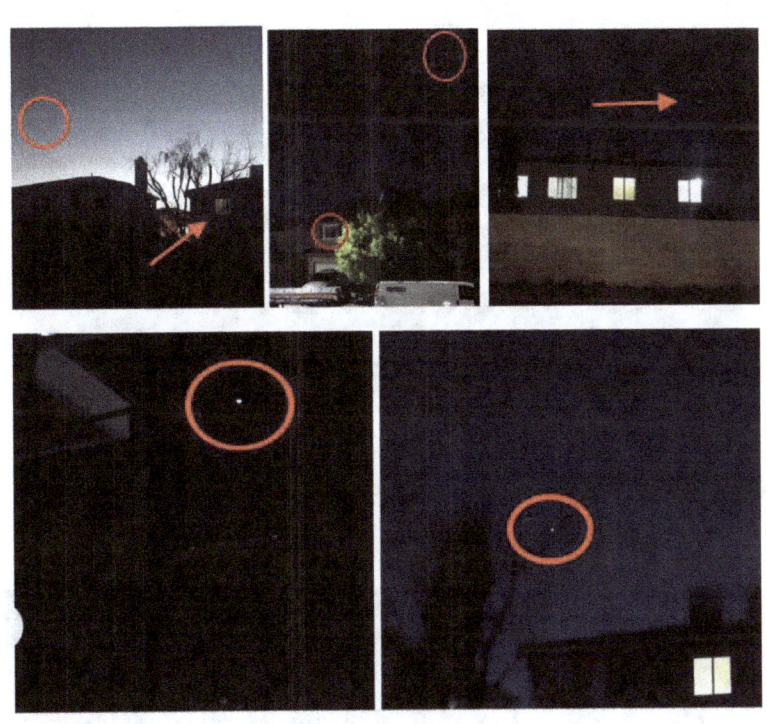

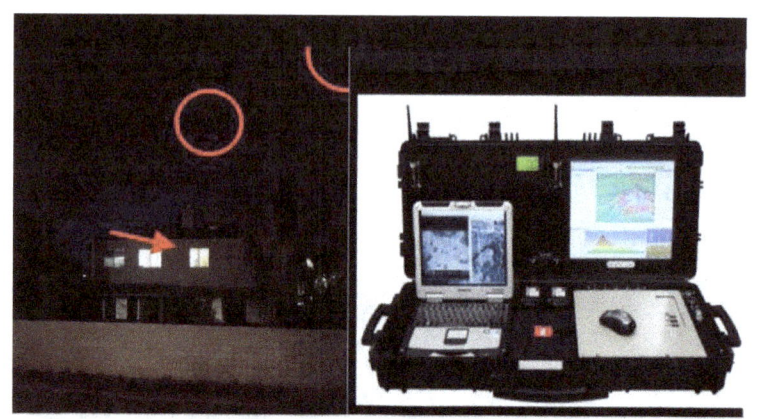

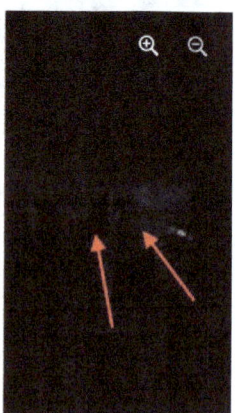

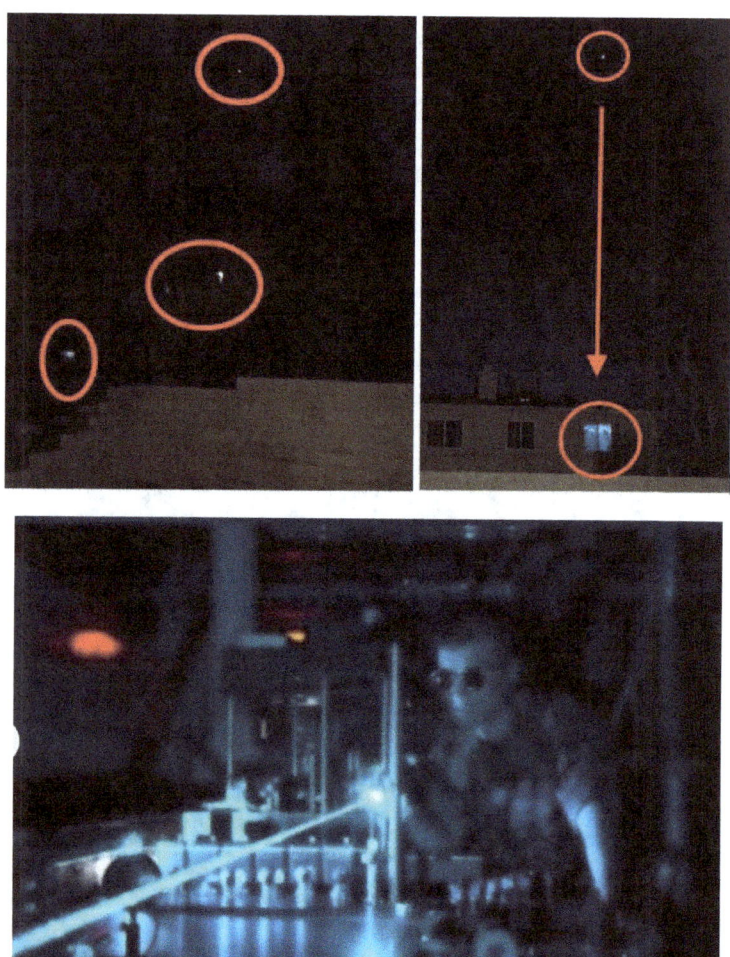

Upstairs you can also see specific equipment used as surveillance technology that, again, when fully deployed emits a lavender or blue illuminated hue color in the room, and also the bright circular light of the Sonic Weapon in images as well.

I leave the TV on to drown out the sonic hummm. Whenever I fall asleep and the TV turns off, they immediately hit me with this Sonic Weapon in the wee hours of the morning confirming the official murderous assignment.

Many nights it was turned on in full force until I woke, having to use the restroom and realized this and turned the TV back on to drown out the hum.

The days of labeling people used as human guinea pigs as mentally ill are over for them. If I and many others, have our way, courageously exposing this program against the odds.

I join many who stand on the truth and are unafraid to publish it! My goal and that of many is to save lives with knowledge and awareness while this hideous hidden official high-tech targeting program continues to destroy, not only men and women deemed human guinea pigs but also children and some report part of the high-tech terrorism pets to destroy targets emotionally. Children as excellent guinea pigs considered beneficial when programmed early in life as shown in the linked information below.

Associated Mind Control Technology Renee Pittman You Are Not My Big Brother Blog

Mind Control & the Early Targeting of Children Blog Link

Mind Control and the Early Targeting of Children & How Family Targeting is Brilliantly Orchestrated for Discrediting Whistleblowers

What will be their next move after insinuation of highly credible whistleblowers as being crazy, and thousands among millions being used step forth and the coverup no longer effective? Time will surely tell. There is an understandable desperation to keep this program hidden.

Across the USA many are determined to contribute to saving lives by knowledge and awareness with gratitude to the some targeted over 30 years who helped many understand the truth. For this reason, I am on board and pay it forward.

Silly me, frankly, in spite of it all, I find this life amazing. Who knew that everything in life since a child prepared me for this mission? What a privilege to contribute to saving lives this way!

Those involved in this program are the truly crazy ones with borderline socio and psychopath tendencies created by government sanction and approval for massive, ruthless human experimentation and destruction of anyone.

Many are devastated, and sadly lost hope crushed and broken both mentally, physically, emotionally due to beamed deterioration of their health and psychologically drained.

The cold hard truth is that our government has factual secret programs for psychophysical secret weapon testing on citizens who are horrifically being denied, Constitutional, Civil and basic Human Rights.

Nationwide, communities have become little more than mass human experimentation playgrounds advancing inventions with a foundation for mass control and BCI and AI and electromagnetic mind and body experimentation that is unbelievably inhumane.

When I moved here, after getting the key, the current situation around me was set into motion before I arrived.

Welcome To The Neighborhood!

September 5, 2014

A Gift Bag Of Poop On The Porch

***

It's a Good Thing I Am Not Trying to Win a Ms. Congeniality Contest. I Don't Care!

What I do care about, deeply and to the core of my existence, is bringing hope and hopeful lifesaving awareness to those whose life can be completely destroyed, who become deathly angry and it results in innocent casualties.

As of 2023, and since 2015, the corner house behind remains a primary location for a variety of official silencing efforts, With so many attacks originating from this location, I consider the house an undeniable house of official high-tech horrors. As shown, since moving here in 2014, the multi-agency effort has become massive as of 2024.

As the plot thickens, one thing is sure, and I fearlessly declare, God is so much bigger than this! and official Workers of Iniquity!

I ask God to prepare my mind and heart for whatever they are creating, or what may come then keep moving forward. This program is not around me, or many since childhood to watch any target prosper. That is not what mind control is about.

It is about destroying hopes and dreams. The bigger your hopes, capability and progression toward life goals, the better the challenge for this herd.

House of High-Tech Horrors Operation
(Corner House Behind)

Once while driving by, an enlisted USAF white kid stood in the yard with his camouflage uniform pants sagging and a bandana covering his face in front of the ZCVB behind. He outrageously thought himself a so-called Gangster.to which revealed the minds of those at the helm of this technology over all.

Following is the third house from the corner which was setup as of June of 2023.

The faint blue illuminated room upstairs, top right, of the house is typical of military surveillance technology. They moved here when I blocked their view neglecting an opening in the protective material from this direction. On this night, as I prepared for bed, in preparation for whatever they have in store for me they lay in waiting for ongoing beamed assaults to my head. Of course, military personnel and in this case recruited contractors are fully aware that if they persist with beamed focus to the area that supplies blood to the brain it could create a brain aneurysm.

For years there has been controversy around Project Blue Beam and reported conspiracy theories of its use to duped humanity into a belief of an alien invasion.

Personally, I am not one on the alien bandwagon and don't believe in their existence. What I have learned in 18 years of relentless research, of this technology are individuals at the helm is that they not with green but red-blood. However, as shown in the following images, you can't blame this program for trying the alien tactic which I am not buying. This promoted belief is part of the ongoing. heinous official coverup.

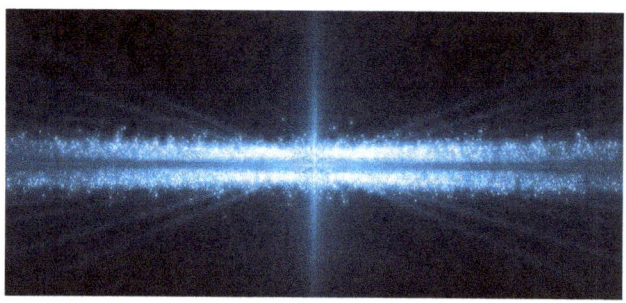

US Shoots Down Mystery Objects, Speculations Over Aliens, 'Project Blue Beam' Grow

Ever since the US shot down mystery objects in February, there have been rumors that the object might be an alien spaceship. However, there has been no official confirmation about the nature of unidentified flying objects, which were shot down, according to a Daily Star report.

American fighter jets shot down four high-altitude objects in airspace above the US and Canada starting from February 4. While the February 4 spy balloon was linked to China, other three objects sighted over Canada, Michigan, and Alaska are yet to be classified.

When I enlarge the image on the next page, captured by my iPhone, I could see imagery of what looks like a fake aliens infused in the beam which I circled in red on the bottom right. This is a room in the corner house behind again captured when I woke under Sonic Weapon assault, this time around 3:00 a.m.

This is an ongoing brainwashing them and it typical of military Psychological Operations designed to foster a belief of alien involvement and use of advanced technology and who is using it.

As shown, DOD Contractors, i.e., Lockheed Martin personnel are heavily involved in this program and as mentioned previously training everyone around me and part of this deception.

Tactics such as this continue today trying to convince people of alien technology when it the brilliance of the human mind and technology that dates as far back, on the record, to Ancient Egypt.

DOD Contractors and the Alien Deception

<u>Unidentified no longer!</u>

Aerospace technology has come a lot farther than most people are aware.

We have been building flying saucers, triangles and the glowing balls of light that shoot through the sky at thousands of miles per hour for over a century. The government has confirmed this though an unexpected agency: The Patent Office.

For further details, check out *The Abridged Original Edition UFO How-To Books*

Again, this is a typical distraction within the official coverup of monstrous human experimentation. The advanced technology being used today is used by clean-shaven, suited and in uniform personnel.

In spite of this truth, they persist in attempts to redirect the focus away from research from agencies like DARPA, the Pentagon's scientific arm, and the use by intel agencies, DOD contractors today police and ongoing military mind control experimentation unifying federal, state and local police forces to aliens.

Renee Pittman Associated Video

A Program so Monstrous, Most Cannot Believe it Exists

https://www.youtube.com/watch?v=t4azoEGbRO8&t=7s

***

Electromagnetic Radiation Effects on Weapons and Energetic Materials

Agency:	Department of Defense
Branch:	Air Force
Program | Phase | Year:	SBIR | Phase I | 2014
Solicitation:	2014.1
Topic Number:	AF141-131

❶ NOTE: The Solicitations and topics listed on this site are copies from the various SBIR agency solicitations and are not necessarily the latest and most up-to-date. For this reason, you should use the agency link listed below which will take you directly to the appropriate agency server where you can read the official version of this solicitation and download the appropriate forms and rules.

The official link for this solicitation is:
http://www.acq.osd.mil/osbp/sbir/solicitations/index.shtml

Description:

Objective:

Perform innovative research in electromagnetic radiation (EMR) effects on energetic materials (EMs). Develop electrical and physicochemical models of EMs to predict safety and energy output under various physical, chemical, and EMR conditions. DESCRIPTION: Ubiquitous radio frequency (RF) and electromagnetic radiation (EMR) usage is an important characteristic of modern military. While EMR offers many benefits, there are disadvantages associated with the EMR-rich environment. One of the disadvantages of EMR is Hazard of Electromagnetic Radiation to Ordnance (HERO). HERO may cause inadvertent activation of a munition. A fundamental and scientific understanding is lacking as to why and how EMR impacts explosives at the molecular level and causes an inadvertent activation of the explosive. Besides the inadvertent initiation of an explosive material, EMR could alter an explosive's performance.

The fundamental mechanisms that take place when an explosive is exposed to EMR needs to be studied, modeled, and validated. This is basic research that brings multiple scientific fields together. The goal of this innovative research is twofold: to develop a physicochemical model, and an electrical model to predict energetic material performance both for safety as well as for energy throughput.

The models developed should incorporate all physical, chemical, temporal, and EMR parameters. The purpose of such models is to help understand the safety features needed in the design of a munition and also to design new materials. The question that needs addressed in this research through the validated models is: What characteristics of EMR and energetic material play a critical role for both safety as well as throughput? PHASE I: Develop theory, and develop first principle physics based physicochemical and electrical models and algorithms for EMs and EMR interaction. Deliverables include a technical report documenting the theory of model development, proof that models and algorithms are relevant, feasible and valid for at least some classes of materials under key parametric conditions of EMR and materials. PHASE II: Enhance and refine the models and algorithms developed in Phase I. Show by analysis and experiments that they are accurate and valid for multiple classes of materials under a number of various physical, chemical and EMR parametric conditions. Deliverables include detailed technical reports, algorithm and model specifics, rationale, and experimentally demonstrated validation evidence for multiple materials under scores of parametric conditions of EMR and material property variations. PHASE III DUAL USE APPLICATIONS: Military applications are munitions safety, and munitions performance characterization. Commercial applications include explosives safety weapons for law enforcement; automobile airbag enhanced safety, reliability, and efficient airbags.

> Torture. 2022;32(1,2):280-290. doi: 10.7146/torture.v32i1-2.132846.

The future is here: Mind control and torture in the digital era

Pau Pérez-Sales [1]

Affiliations + expand
PMID: 35950441 DOI: 10.7146/torture.v32i1-2.132846
Free article

Abstract

Torture, understood as a relationship of domination in which one person breaks the will and impedes the self-determination of another human being, taking control of all aspects of the victims' life and trying to change the core elements of their identity to the perpetrator's interests (Pérez-Sales, 2017), will increasingly come to be linked to new technologies, artificial intelligence, the use of media and internet, and to new forms of lethal and non-lethal weapons. The author reviews the implications of modern technology for the contemporary fight against torture and some of the emerging civil society initiatives that aim to face them.

Keywords: Torture, Non-Lethal weapons, Neuro-warfare, Nanotechnologies, Mind control. Surveillance Methods, Neuro-ethics, Cognitive Liberty.

There Is Nothing New Under The Sun!

Discrediting Is Key Which Allows This Program To Continue Due To Disbelief

***

Military Psychological Warfare Rewind

Psychological warfare (PSYWAR), or the basic aspects of modern **psychological operations (PsyOp)**, have been known by many other names or terms, including Military Information

Support Operations (MISO), PsyOps, political warfare, "Hearts and Minds", and propaganda. The term is used "to denote any action which is practiced mainly by psychological methods with the aim of evoking a planned psychological reaction in other people".

Various techniques are used, and are aimed at influencing a targeted audience's morals, value system, perceptions, belief-system, emotions, motives, reasoning, or behavior.

It is used to induce confessions or reinforce attitudes and behaviors favorable to the originator's objectives, and are sometimes combined with black operations or false flag tactics. It is also used to destroy the morale of enemies through tactics that aim to depress troops' psychological states.

Target audiences can be governments, organizations, groups, and individuals, and is not just limited to soldiers. Civilians of foreign territories can also be targeted by technology and media so as to cause an effect on the government of their country.[5]

***

Gaslighting: A Precursor For The Mental Illness Tag

[Excerpt]

Gaslighting is a form of psychological abuse where a person causes someone to question their sanity, memories, or perception of reality. People who experience gaslighting may feel confused, anxious, or unable to trust themselves.

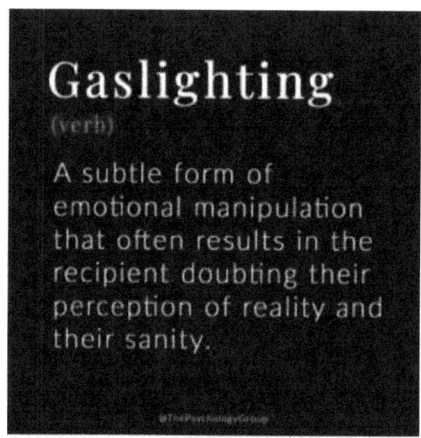

In the grand scheme, I guess no one can blame this corrupt, official, high-tech targeting program for trying and trying anything they can to keep the truth under the lid and it makes complete sense. This program is on a mission of training and ruthless nonconsensual human experimentation for coming generations.

NSA Whistleblower Powerhouses William Binney and Kirk Wiebe Stand Up to Support "Targeted Individuals" Worldwide

[Excerpt]

In a breakthrough historic and supportive move on a Talk Shoe podcast call-in show, and a first in terms of notable public figures publicly acknowledging covert targeting with electromagnetic and neuro-weapons as real, NSA whistleblower power houses **William Binney** and **Kirk Wiebe** stepped forward to announce their support, concern for, and distinct plans to assist the entire community of Targeted Individuals in the United States and worldwide.

***

Confirmation that Targeting Operations Hide in Communities Testing Electromagnetic Weapons on Anyone

Renee Pittman You Tube Video

https://youtu.be/1HzszdqhOcY

My revealing homes in my community where high-tech operations are setup has become a major thorn in their side, however, they leave no choice joining many across our nation. Frankly when exposing this type of official corruption, the mission is focused on permanent silencing.

I will not sit inside my house, as a sitting duck and targeted by relentless beamed harassment. In spite of this, it must be clearly understood, beyond a shadow of a doubt, that the best fighter is never angry. This is secret weapon.

Many have been conditioned to be afraid to speak up about what is happening to them, and the mental illness, manifestation plays a major role in this. *I say, Just don't call me late for lunch!*

With many revealing mind invasive technology by agencies we are programmed to trust, the only reason people are still standing in the face of powerful, high-tech corruption is likely due to exposure.

Many people have been bogusly placed on a Watch-list, then marked for experimentation for, again, and various reasons. The watch list approves official human experimentation.

All you really need to do in order to be tagged as a suspicious character, flagged for surveillance, and eventually placed on a government watch list is live in the United States.

This is how easy it is to run afoul of the government's many red flags.

In fact, all you need to do these days to end up on a government watch list or be subjected to heightened scrutiny is use certain trigger words (like cloud, pork and pirates), surf the internet, communicate using a cell phone, limp or stutter, drive a car, stay at a hotel, attend a political rally, express yourself on social media, appear mentally ill, serve in the military, disagree with a law enforcement official, call in sick to work, purchase materials at a hardware store, take flying or boating lessons, appear suspicious, appear confused or nervous, fidget or whistle or smell bad, be seen in public waving a toy gun or anything remotely resembling a gun (such as a water nozzle or a remote control or a walking cane), stare at a police officer, question government authority, or appear to be pro-gun or pro-freedom.

We're all presumed guilty until proven innocent now.

***

We're all presumed guilty until proven innocent now.

As many come forth across the USA reporting being placed on a Government Watch-List the numbers are growing into millions. Sure, the average person would think, what is the issue of being watched as part the War on Terrorism resulting from the nationwide trauma of 9/11?

The main issue widely reported and intentionally hidden is that people are fraudulently being placed on this list which allows military technology use. In reality this is connected to ongoing massive human experimentation programs by the government, today unifying federal, state and local police departments with military technology and military personnel.

They are using highly advanced, patented, psychophysical, mind invasive technologies, that include bioweapons, and a vast array of patents known as psychological electronic (Psychotronic), psychophysical technologies that can be, and are being used to not only invade the mind and body but tamper with both. At the helm are what people describe as evil, official and horrifically unscrupulous individuals attempting to use this technology to create the goal they seek testing the technology to it limits and beyond.

As a result, the "Watch-List" has become a haven for nonconsensual mind control and behavioural modification experiments and, for example, DARPA's "Brain Initiative". This is the foundation for scientific advancement, nationwide and globally, for the Neuroscience explosion, mass population hive mind technological testing, and advancing Brain Computer Interface (BCI) and Artificial Intelligence (A.I.).

Below the excerpt confirms that Lockheed Martin is factually heavily involved in field training on electromagnetic weapon systems, devices, drones and satellite connected technology across the USA and globally.

LOCKHEED MARTIN INVOLVEMENT IN ILLEGAL TARGETING

One of my confidential sources in fact, passed on the following assessment of Lockheed Martin and their involvement in the secret and illegal field-testing of such weaponry upon innocent people both in the USA and abroad:

> "Lockheed Martin, the world's largest defense contractor and the prime cyber security and information technology supplier to the Federal government, coordinates the communications and trains the "team leaders" of a nation-wide, Gestapo-like apparatus, which has tentacles into every security and law enforcement agency in the nation, including state and local police and 72 regional 'fusion centers' administered by the U.S. Department of Homeland Security. According to company literature, Lockheed Martin has operations in 46 of the 50 states.
>
> Lockheed Martin also has operational control over a U.S. government microwave radio frequency weapon system, deployed on cell tower masts throughout the U.S., that is being used to silently torture, impair, subjugate and electronically incarcerate so-called "targeted individuals." The nexus of this American torture matrix appears to be Lockheed Martin's Mission and Combat Support Solutions central command center in Morristown, PA., which employs several thousand workers. The defense contractor's global headquarters is located in Bethesda, MD., just outside the nation's capital.
>
> Lockheed Martin, under contract to U.S. Security and intelligence agencies and commands, also conducts warrantless surveillance of the telecommunications of targeted persons, and routinely censors and tampers with the content of their communications, as this [CENSORED] has documented in a series of recent [CENSORED].
>
> Thousands of Americans, including this [CENSORED] have [CENSORED] reported being the victims of silent electromagnetic

Confirmation letter, revealing Lockheed Martin field training of weapon systems of personnel inside the USA and globally posted on social media by targeted whistleblower Karen Melton-Steward, 28-year veteran of the National Security Agency assault and community-based home intrusions, and even the poisoning of food, water and air. Federal and local law enforcement routinely dismiss their reports as the product of delusions or mental illness, and refuse to investigate their complaints.

These 'targets' and their families have been physically harmed and financially destroyed as a result of tax-payer funded " psychological operations", police-protected community "stalking" harassment and malicious vandalism, and other covert programs of personal destruction, including government-assisted financial sabotage. Many appear to have been targeted as a result of their politics; their activism or corporate whistleblowing activities; their ethnic background; or as a result of score-settling vendettas by persons in positions of power – in government and the private sector."

* Note that this fits perfectly with Mrs. Stewart's claims that she was stalked and harassed first in 2006-2009 for reporting wrong-doing by NSA management to the NSA IG, then again beginning in April 2015 after subpoenaing corroborating information against retired NSA executive Eric Hagemann, and then electronically harassed beginning in late Nov 2015, after a dust up with 16th Deputy Director of NSA, Bill Black Jr. in regard to her firsthand knowledge exposing of 9/11 as entirely preventable, a false flag operation, on social media.

An additional report from a protected source, on Lockheed Martin indicates that this corporation is indeed involved in criminal activity under the protection of the U.S. government intelligence community and military:

LOCKHEED MARTIN:

"On the surface, Lockheed Martin appears to be a (sic) above board company. Below the surface it is the real story of how they are the masterminds of gang stalking and etc. Their teams of nameless Black Ops are specially trained in these techniques. One (sic) you make these lists as a TI you are in the world of hell. No one has ever gotten

National Fusion Center Connection and Approval

As the veil lifts, with focus on is the actual leadership, all finger's point to nationwide DHS and FBI Fusion Centers which sprung up Post 9/11. In Southern California, the culprit is the Joint Resource Intelligence Center (JRIC) in Norwalk, California which serves as a nationwide model with all connected to Military Intelligence official civilian experimentation programs.

This JRIC is comprised of the following cooperating agencies:

Federal Bureau of Investigation, Department of Homeland Security, State of California, Department of Justice, Office of Homeland Security, Los Angeles County Sheriff's Department, Los Angeles Police Department, Long Beach Police Department, Los Angeles Airport Police Department. There are 80 Fusion Centers nationwide and by some accounts growing into over a hundred. However, these figures do not include military centers from bases across the USA and modernized police station operations.

Department of Homeland Security Fusion Joint Resource Intelligence Centers Grand Opening in California

U.S. and Southern California law enforcement will have a grand opening of the Joint Regional Intelligence Center. If you have any questions regarding this event, please call Media Relations Section at 213-485-3586. Parking for media vehicles and crew will be available directly in front of the building (alongside Imperial Highway). The building is privately owned and its management request that no live shots or exterior photos of the building be taken. Photos are limited to the interior of the JRIC/JDIG space only. Media will be admitted to the event beginning at 10:00 a.m.

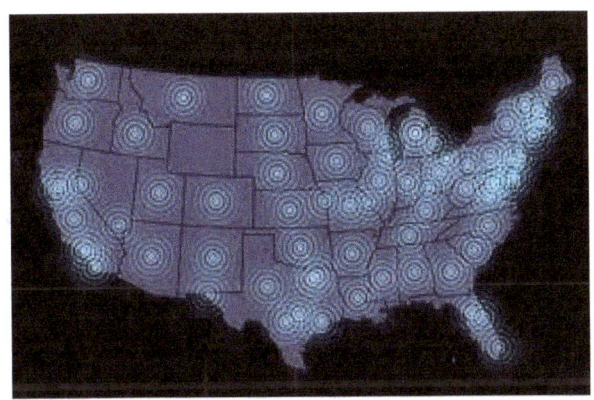

The ACLU link below details intel agency, fusion centers and this unified military connection. Used as an example is also a Department of the Navy (DoN) image revealing how satellites and drones pass information through the fusion. One thing is certain, these agencies will never confirm the connection to military electromagnetic, psychophysical, psychotronic and mind invasive technologies of various types including officially patented mind reading and subliminal influence technologies

Biometrics is the New Paradigm

Description

Biometrics Enabling Capability (BEC) (formerly the Biometric Enterprise Core Capability [BECC]), using an Enterprise System-of-Systems architecture, will serve as DoD's biometric repository, enabling multi- modal matching, storing, and sharing in support of identity superiority across the department.

Understanding Psychological Electronic, Psychotronic Military Technologies and The Influence of Microwaves On Living Creatures

Behavior In 1975, a neuropsychologist Don R. Justesen, the director of Laboratories of Experimental Neuropsychology at Veterans Administration Hospital in Kansas City, unwittingly leaked National Security Information. He published an article in "American Psychologist" on the influence of microwaves on living creatures' behavior.

In the article he quoted the results of an experiment described to him by his colleague, Joseph C. Sharp, who was working on Pandora, a secret project of the American Navy.

Don R. Justesen wrote in his article:

By radiating themselves with these 'voice modulated' microwaves, Sharp and Grove were readily able to hear, identify, and distinguish among the 9 words. The sounds heard were not unlike those emitted by persons with artificial larynxes (pg. 396).

That this system was later brought to perfection is proved by the document which appeared on the website of the U.S. Environmental Protection Agency in 1997, where its Office of Research and Development presented the Department of Defense's project: Communicating Via the Microwave Auditory Effect.

In the description it said:

An innovative and revolutionary technology is described that offers a low-probability-of-intercept radiofrequency for use in (RF) communications The feasibility of the concept has been established using both a low intensity laboratory system and a high-power RF transmitter. Numerous military applications exist in areas of search and rescue, security and special operations.

Military And U.S. Law Enforcement Experimentation Are Connected To The Biometric Database

Does anyone remember that it used to be considered a conspiracy theory to warn people about biometric databases? Well, now that it is an accepted reality, we should be looking even farther down the slippery slope for new signposts indicating even greater plans to track, trace and database all human beings.

Apparently, it is neither sufficient — nor *efficient* — to have the myriad databases currently at the disposal of agencies like the Department of Defense, FBI or Homeland Security. According to a release from *Biometric Update*. com these agencies currently use different computer languages that have difficulty communicating with each other as seamlessly as the U.S. government desires.

***

Everyone's biometrics were and continue be downloaded into a supercomputer system, drones also have the capability for capturing biometrics. This agenda evolved through the government's Total Information Awareness Program. It, in turn, opened the door for anyone to be effortlessly targeted and tracked by advanced technologies, DNA, iris, facial and voice recognition and gait, and anywhere on the face of the Earth instantly.

***

Our biometric signature is our unique brainwave fingerprint. DNA is also part of biometric stored information.

It's an auditory effect that's created by military scientists who manipulated the air with lasers — and it's the Pentagon's most interesting idea for stopping people charging checkpoints, or just scaring the crap out of them.

The U.S. military's Joint Non-Lethal Weapons Program, or JNLWD, factually created weapons that alters atoms to literally create words from thin air. It's called the Laser-Induced Plasma Effect and carrying words through the air. For its use, biometric tracking and monitoring of people is combined and this can happen any place on the face of the Earth.

Total Information Awareness Program And Your Biometric Signature

The human identification at a distance (Human ID) project developed automated biometric identification technologies to detect, recognize and identify humans at great distances for force protection, crime prevention, and homeland security/defense purposes

***

Brain Computer Interface and Neuroweapons

Abstract

Brain reading technologies are rapidly being developed in a number of neuroscience fields. These technologies can record, process, and decode neural signals. This has been described as 'mind reading technology' in some instances, especially in popular media. Should the public at large, be concerned about this kind of technology? Can it really read minds? Concerns about mind-reading might include the thought that, in having one's mind open to view, the possibility for free deliberation, and for self-conception, are eroded where one isn't at liberty to privately mull things over. Themes including privacy, cognitive liberty, and self-conception and expression appear to be areas of vital ethical concern. Overall, this article explores whether brain reading technologies are really mind reading technologies. If they are, ethical ways to deal with them must be developed. If they are not, researchers and technology developers need to find ways to describe them more accurately, in order to dispel unwarranted concerns and address appropriately those that are warranted.

***

Neuroweapons: No Chip Or Nanotechnology Required

While these advances can be greatly beneficial for individuals and society, they can also be misused and create unprecedented threats to the freedom of the mind and to the individuals' capacity to freely govern their behavior…"

***

The personnel at the helm of this technology, 100%, are official military and police, or those who were DOD Contractor recruited, i.e., Lockheed Martin, Northrup, Raytheon, etc., and who qualify after lengthy, Top-Secret Clearance investigation becoming employed DOD Contractors.

Renee Pittman Website *Technology Approval*

Reference Link

bigbrotherwatchingus.com

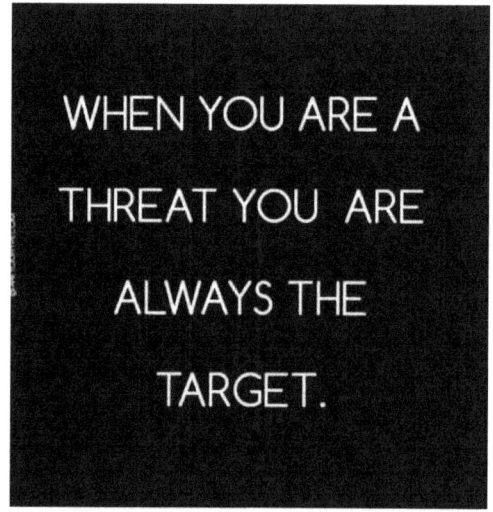

Some people are motivated to do evil to gain self-esteem, to live up to a grandiose self-image. In these situations, vindictive acts can be weapons used to put other people down. However, many evil acts are the result of idealistic fanaticism…

Running this program today, sadly, are official personnel whom all our lives we are programmed to trust, honor and respect believing they will always do the right thing and act honorably.

With horrific suffering reported nationwide, in ongoing official human experimentation programs targeting men, women and children, I and many ask how can they be so vicious and inhumane? What personal satisfaction do they gain from destroying lives while hiding behind these systems?

With some, they perceive targets as some sort of twisted challenge, of which the target's demise validates a horrific hunger. They don't take to kindly when they realized, many targeted, will not lay down and be walked over, and this is especially true, when and you begin telling what they are really doing. This is when things get ugly.

One thing is certain, they do not care about the truth of this being factual massive human experimentation, with some before they were born, and ongoing for decades.

This truth reveals them as inhumane, ruthless individuals devoid of care, and nowhere close to heroes.

Somehow, in their official programming they have compartmentalized the truth, even as they sit before this technology manipulating and influencing thousands, to feed some type of thirst within. The answer lies in the heinous, covert, destruction of lives as if lives do not matter proving evil.

And, when they realize many will not allow ourselves to be used, and do not fear them, this tampers with an ego driven psyche, and grandiose perceptions they have of themselves. How dare you? They seem to think. We can kill you!

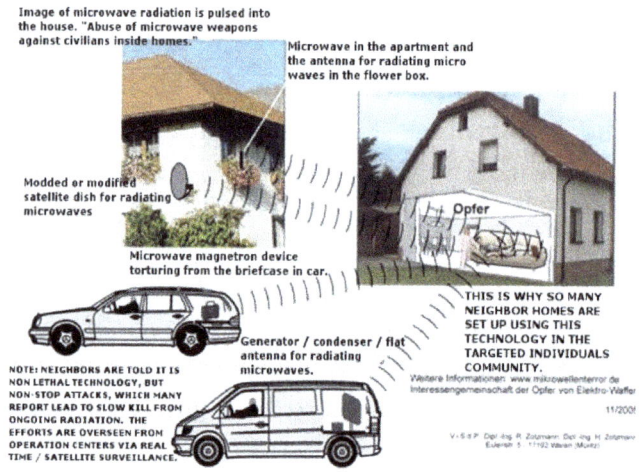

Patent: Multifunctional Radio Frequency Directed Energy antenna System United States Patent – 7629918 B2

ABSTRACT: A RFDE system includes an RFDE transmitter and at least one RFDE antenna. The RFDE transmitter and antenna direct high-power electromagnetic energy towards a target sufficient to cause high energy damage or disruption of the target. The RFDE system further includes a targeting system for locating the target. The targeting system includes a radar transmitter and at least one radar antenna for transmitting and receiving electromagnetic energy to locate the target. The RFDE system also includes an antenna pointing system for aiming the at least one RFDE antenna at the target based on the location of the target as ascertained by the targeting system.

The garage next door (North) is a primary setup for initially USAF military personnel beamed weapon assaults into my bedroom while I sleep closest to my bedroom.

On this day the person operating in the garage setup puts his USAF boots outside to air out.

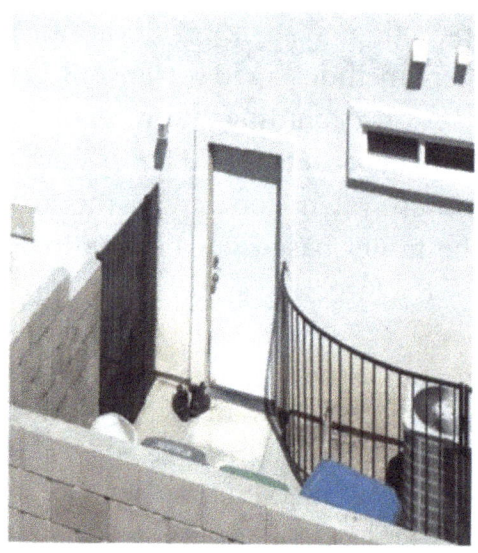

The weaponized military drone operation Panasonic Toughbook, shown in the following image, is the portable system used to operate the weaponized drones and strategically placed antennas.

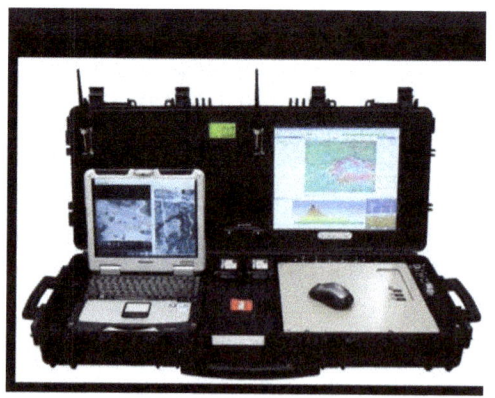

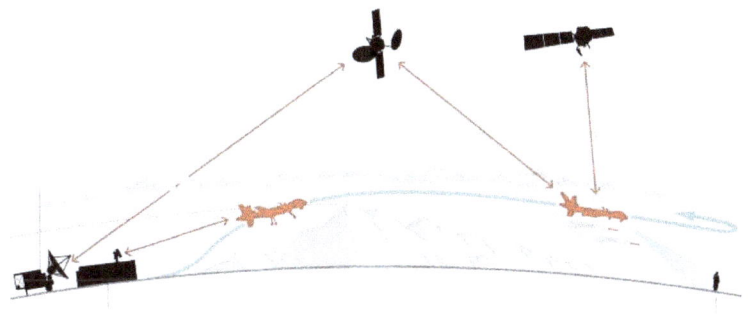

How Drones Are Controlled

Pilots rely on satellites to track drones

From takeoff until it leaves the line of sight, the drone is controlled with a direct data link from a ground-control station.

***

Military X-ray drones see through walls

A pair of drones can use Wi-Fi signals to see through walls.

Researchers at UC Santa Barbara were able to create three-dimensional images of the objects behind a brick wall in a series of experiments with the drones.

The two flying machines work in tandem. In the demonstration, they fly around a four-sided brick building. One drone transmits a continuous Wi-Fi signal, while the other, on the opposite side of the house, measures its power after it passes through.

By circumnavigating the house several times, the drones can generate high-resolution, accurate 3D images of the objects inside.

By providing technology with the ability to see inside homes, to the community, it is clear that this program wants to violate targets in every way possible way and this part of the psychology of the 24/7 assaults. Women suffer from the invasion of privacy when showering and dressing which can be emotionally devastating.

There was once a Constitutional, Fourth Amendment Right for protection inside private residences which that no longer exists as a direct result of modernized aerial surveillance .

Detailed Link

https://epic.org/issues/surveillance-oversight/aerial-surveillance/

NOTE: While I am drone tracked and monitored around town there are two drones working in sync, a larger drone connected to a small drone. You can see both lights in the night's sky. The video linked was captured during the day.

Below is a You Tube video link captured of drone tracking of Renee Pittman while running errands entitled: *Drone tracking and TARGETING of Human Rights Advocate* **Renee Pittman**

You Tube Video

https://www.youtube.com/shorts/lpTVYeql0e4

Similar videos of personal drone tracking can be seen on:

You Tube Channel

Renee Pittman Books

@humanrightsadvocatereneepi7422

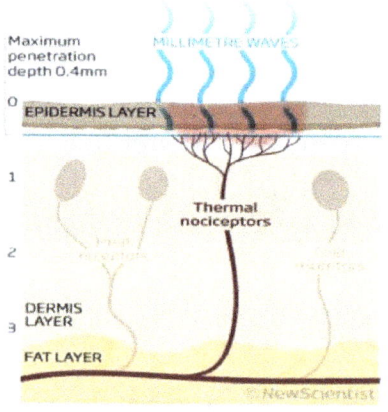

The Military Active Denial System Pain Ray is Today's Official Tortuous Weapon of Choice

The Active Denial System (ADS) is a non-lethal directed-energy weapon developed by the U.S. military, designed for area denial, perimeter security and crowd control. Informally, the weapon is also called the heat ray since it works by heating the surface of targets, such as the skin of targeted human beings.

The earliest reports of the U.S. military beginning research into the use of lasers in combat is documented to be the late 1950s. Some of the earlier versions were perfected by 1973 in research and testing efforts at the United States Air Force Laboratory at Kirtland Air Force Base in New Mexico. They were the tactical laser, the Mid-Infrared Advanced Chemical Laser (MIRACL), the megawatt deuterium fluoride

(DF laser built by DOD contractor TRW) which was tested against aerial targets. This was followed by the first of its kind, a chemical oxygen iodine laser (COIL).

A wide range of lasers have been developed since then which includes free-electronic lasers, and solid-state lasers. Electromagnetic weapons offer the advantage of scalability — from microwaves that heat the skin to make the target extremely uncomfortable with high intensity pain but without injury, to high-power electromagnetic weapons that can destroy an enemy ballistic missile in flight.

It should be noted that by numerous reports across the USA, that microwave weapons if used periodically to heat the skin, are actually harmless, however, people reporting horrendous abuse and suffering from ongoing beamed assaults, used for control and subjugation, reveal that, gradually with repetitive use on focused areas of the body that ultimately become deadly due to repetitive and consistent heating resulting tissue death necrosis.

The use of weaponized Predator drones operated by portable military computerized systems today is what is being widely reported as tested today on civilians. The aircraft can employ two laser-guided missiles, Air to Ground Missile-114 Hellfire, that possess highly accurate, low-collateral damage, and anti-armor, anti-personnel engagement capabilities.

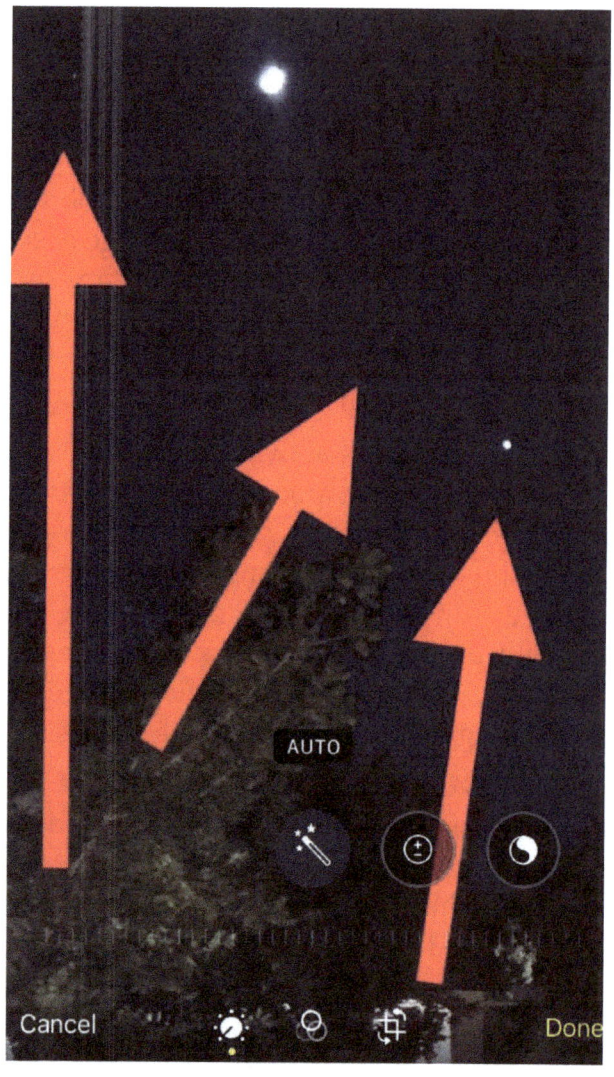

Following is an images of the house directly behind me, operating two drones behind my house. Again, the violet and bluish illuminated rooms are a tell-tale sign of military surveillance equipment used by all. On this night, the beamed tortured me viciously until early morning.

Targeting Setup in House Directly Behind

This program has money to spare and to give away originating from the Government's DOD billion dollars budget! What neighbor could turn down this type of lucrative offer?

Intelligence Budget Data

The Fiscal Year 2023 budget appropriation included $71.7 billion for the National Intelligence Program, and $27.9 billion for the Military Intelligence Program.

I watched military personnel setting up at the house behind through my window one night. After the system was fired up, and the monitor on, they saw me watching and closed the curtains.

The drone empowered antenna today in house is adjusted for a clear line of sight each night inside my bedroom from behind. Houses, front, back, and on both sides of targets are primary locations for beamed weapon assault setups.

The two following images are the corner house behind, as I explained, it is a major location for everyone to train. Note also the blue beam. In the image below is the technology when it is not active showing a small vertical lamp.

The Sonic Weapon is a weapon of choice from this specific location, however, I have seen two other houses set up with the typically two or three lights that are illuminated in windows.

Here we Find the Satellite Drone Blue Beam

Corner House Behind - The Tiny Image in the Oval Circle is Blue Light Device That Illuminates Rooms When on and creates the Violet or Blue Colors

***

Another Aerial Blue Beam In Backyard Deception

This book threatens to demolish that faith. Because here Mark Pilkington sets out to prove that the U.S. government really has been conducting a top-secret UFO conspiracy - only one designed not to hide UFOs but publicize them, fueling and even creating the major UFO myths. Flying saucers, alien abductions, crash-landed spacecraft, secret underground bases in New Mexico - they were all created by the U.S. government.

***

The strategic coverup and alien connection to me is downright pathetic. For example, in the images following you again see the house, directly up and running, however when I enlarged the satellite to drone beamed imagery on my iPhone picture, you ridiculously see what I believe are aliens infused images. This again is part of high-tech Gaslighting.

Following are two top images reveal the typical violet illumination of the rooms by surveillance equipment at the top of the window. Note that in the bottom left image that has the second arrow, the bottom arrow pointing to what looks like mist. When I enlarged the mist, as shown on the bottom right after enlarged, the image appears to be also infused with what looks like aliens around or a scene from a movie in the background.

I have never seen *Lord of the Rings*. However, in 2008, a beamed hologram was beamed into my apartment, from the same official operation, then set up in the apartment next door. I heard a man I met earlier that day, involved in this program, call out my name around 4:00 a.m. to wake me for the show. When I did, there stood what looked like an alien thingy. I didn't believe it then and do not believe it now! Discrediting people as hallucinating is a major part of the official cover, using beamed images is designed to use the mental illness tag.

When I did a Google search, after enlarging the mist image, bottom right, I realized that the boy shown sitting in what appears to be in a circle with other creatures could be the image of the boy in the movie, *Lord of the Rings*.

Young Sheldon In Lord Of The Rings?

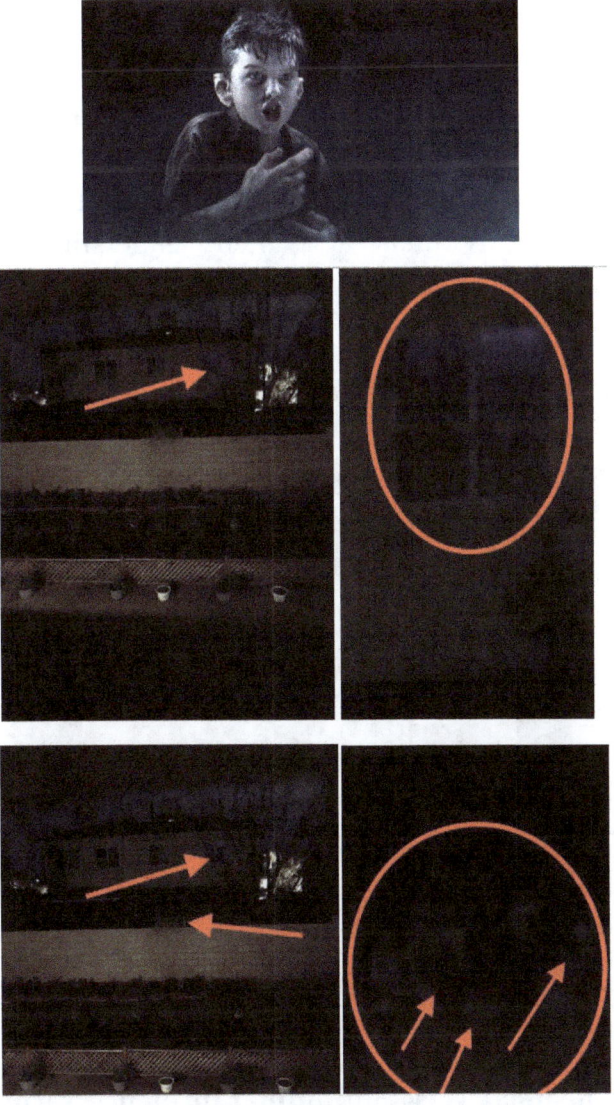

When the mist above, second arrow below left side, is enlarged, it reveals more fake Lockheed / military deception in the scene on the right. Again, this is all part of the alien coverup for military human experimentation and use of mind invasive technology nationwide.

On a personal level, I join many who completely understand that this program is a horrific lie, even the alien farce, and it appears enforced by mentally disturbed, twisted, ego driven individuals and their leadership that have absolutely no regard for human lives and want to use this technology without exposure of what they are really doing.

Below is an example of how drones follow target's around town from one of the established remote locations and also having the ability to subliminally influence people and anyone around targets negatively.

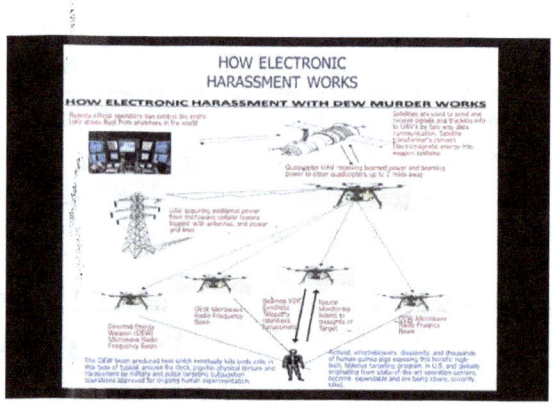

Drone Beamed Mass Subliminal Influence Technology
Subliminal Influence Technology

Subliminal Definition

Subliminal stimuli are messages or images that are perceived by the brain, but only unconsciously. A subliminal message is a message that is not consciously processed by the brain, but still yields a reaction. Subliminal are the messages or the pieces of information that make up the subliminal message. A good example of a subliminal message would be the arrow from A to Z in the Jeff Bezo logo, or the arrow found in the FedEx logo. The main characteristics of a subliminal message are that it is hidden and the unconscious mind is able to perceive it. However, a subliminal message cannot be so hidden that the brain cannot perceive it or the subliminal message will lose its effectiveness. Subliminal are the messages sent with an auditory or visual broadcast below conscious processing. **Subliminal stimuli** are very short auditory or hidden, yet subtle visual messages. Subliminal messages are based on subtle messages.

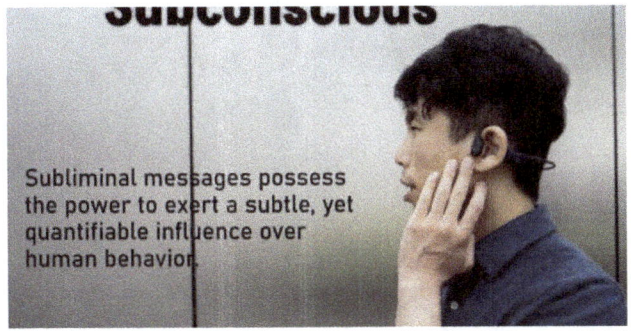

Subliminal Influence Patent Example

Abstract

A computer-readable medium contains software code that, when executed on a computer causes the computer to implement a method for exposing subliminal messages. The method comprises creating a new object, covering a whole or a part of a subliminal message, and uncovering the whole or the part of the subliminal message.

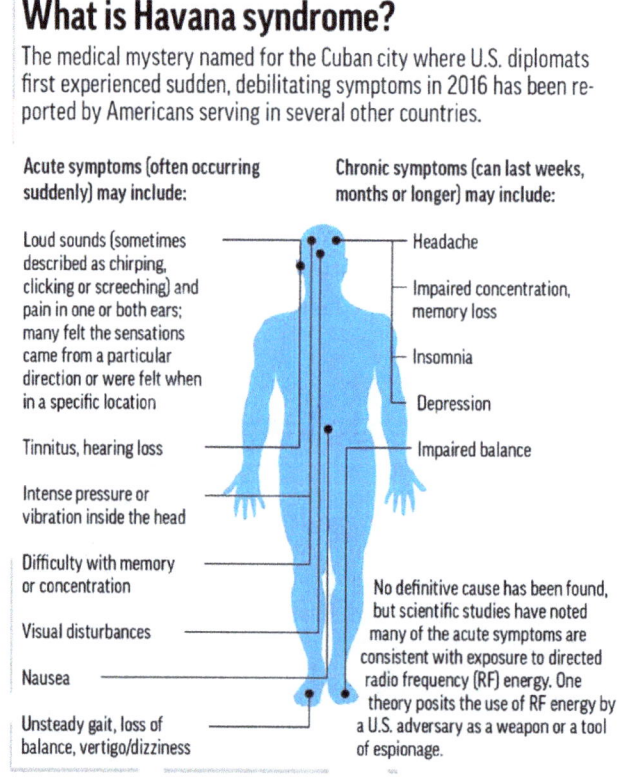

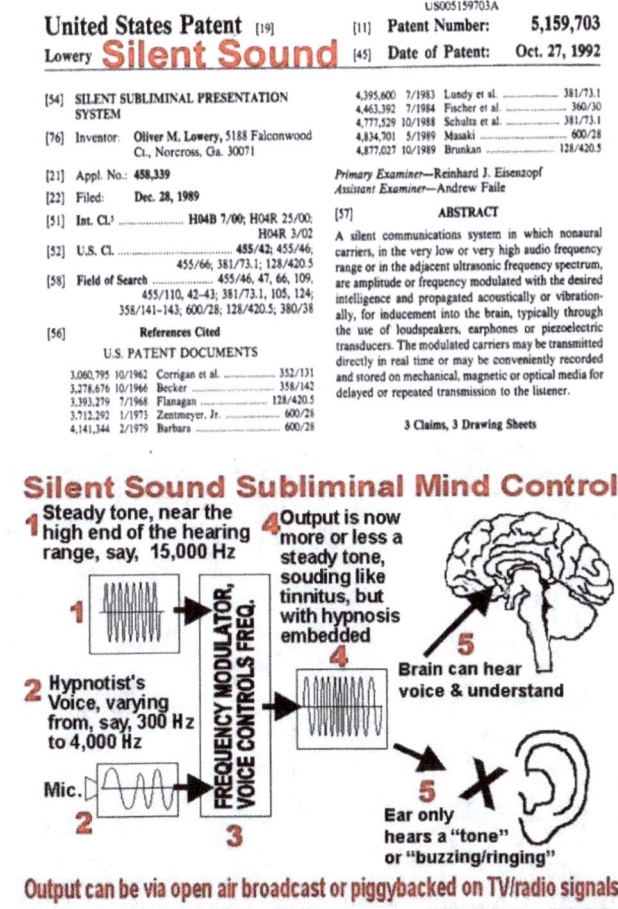

***

With me they hope to create hate directed at me in the public using business that I frequent as pawns accomplished by military advanced technology on drone beamed into their heads. Believe it or not, using the beamed influence technology that mimics any voice people think the subvocal thought materializing as negative thoughts towards the target that suddenly materialized in their heads are their own.

Perfect strangers, while standing in a checkout line, who look as me as if in a trance and have said, out of the blue, *She's gone* or verbalize various other official beamed threats from those at the helm.

Here again is another house across the street where I have seen military personnel using bedrooms upstairs

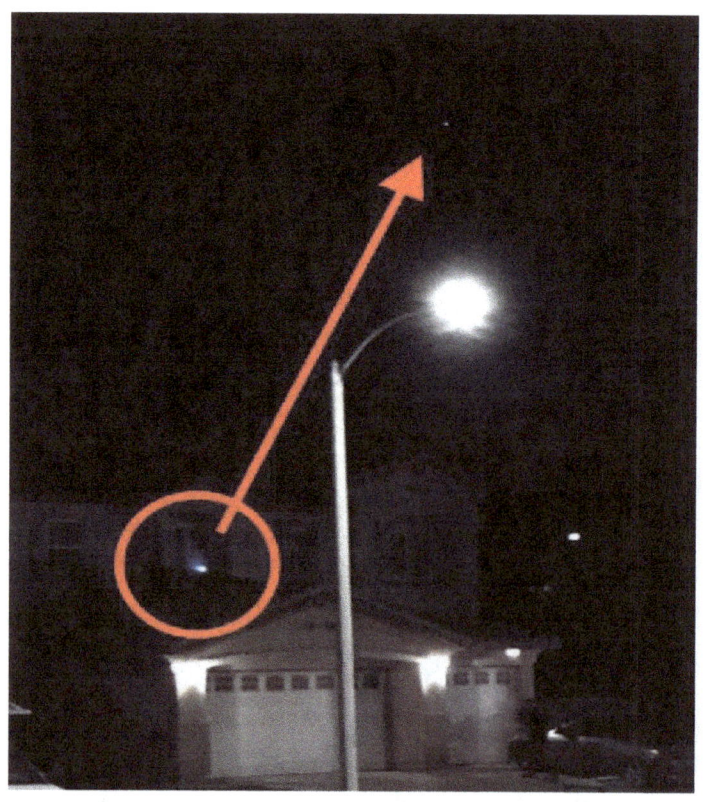

Following is an example of the drone again at a USAF person's house, operated from an upstairs bedroom connected to a blue antenna set up in my next-door neighbor's truck on my south side. This is a bio-coded system. When I am near it began to flash.

Following on another night, at the USAF home, this same house has switched rooms, revealed by the blue illuminated room with a drone over my house and beam assaulting me inside my house. A major focus from this location is my breast, knees and abdomen.

USAF Personnel House Across the Street

Example of Car Parked in Front of My House with Remotely Guided Weaponized Antenna used by USAF

Corner House Behind Sonic Weapon Woke Up

The image, bottom, on the previous page is the stationary drone next door, South Where I Have Seen Federal Agents

***

Lockheed Martin John Norseen Historic was a key player in Brain Mapping and thought control . John Norseen was an American weapons designer working on what today would be referred to as a Neuroweapons, he was also a lecturer at George Washington University in DC. When he was employed by Lockheed-Martin in the 1990s and early 2000s, the concept of neuroweapon was not widely known outside the deepest of black operation funded military and defense sectors. Even, today, the development of such weapons is a highly classified and compartmentalized affair.

His thesis at the War College was on applying Neuroscience research to anti-terrorism investigations. The first article of John Norseen dates to a Newsweek article from2001 he was interviewed for under the title, 'Reading your mind and injecting smart thoughts 'in which he talks of being able to read terrorist suspects thoughts remotely.

In a subsequent article from 2001 in the Washington Times 'NASA plans to read terrorist's minds at airports', Norseen notes: "Space technology would be adapted to receive and analyse brain-wave and heart beat patterns, then feed that data into computerized programs 'to detect passengers who potentially might pose a threat, 'according to briefing documents obtained by the Washington Times. NASA wants to use 'non-invasive neuro-electric sensors, 'imbedded in gates, to collect tiny electric signals that all brains and hearts transmit.

Below the main DOD Contracted trainer arriving for work across the street in the front, left image.

Later he is setup in the wee hours of the morning in the corner house behind. He is shown sitting back to back at weaponized computer systems to one the military personnel he is training who moved into neighborhood.

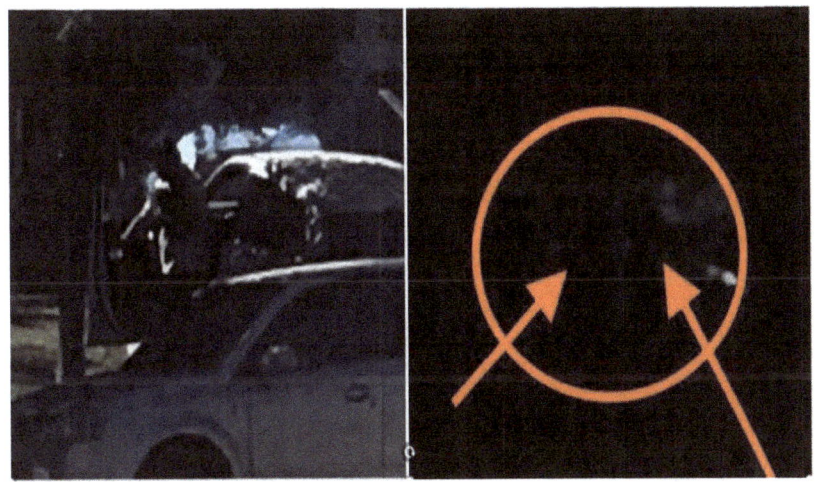

Lockheed History of Brain Mapping

Mapping human brain functions "...Electromagnetic pulses would trigger the release of the brain's own neurotransmitters to fight off disease, enhance learning, or alter the mind's visual images, creating what Norseen has dubbed "synthetic reality." The key is finding "brain prints." "Think of your hand touching a mirror," explains Norseen. "It leaves a fingerprint."

Bio Fusion would reveal the fingerprints of the brain by using mathematical models. Just like you can find one person in a million through fingerprints, he says, you can find one thought in a million. It sounds crazy, but Uncle Sam is listening. The National Aeronautics and Space Administration, the Defense Advanced Research Projects Agency, and the Army's National Ground Intelligence Center have all awarded small basic research contracts to Norseen, who works for Lockheed Martin's Intelligent Systems

Division...By viewing a brain scan recorded by a magnetic resonance imaging (MRI) machine, scientists can tell what the person was doing at the time of the recording – say, reading or writing. Emotions from love to hate can be recognized from the brain's electrical activity. Thought police.

***

Following is the second trainer. He parked in front of my house on this one day, while work was done at the main house they, military and the trainer have leased and my security cam captured him moving cars around.

Later that night, is an image, bottom revealing him setup and sitting at the weaponized deadly beam Sonic Weapon system and using the corner house behind. Again, in this case, it is a blessing that I get up to use the restroom during the early morning hours when they strike. There is only one goal for targeting when a person is sleeping, obviously it is an attempted strategic, slow-kill.

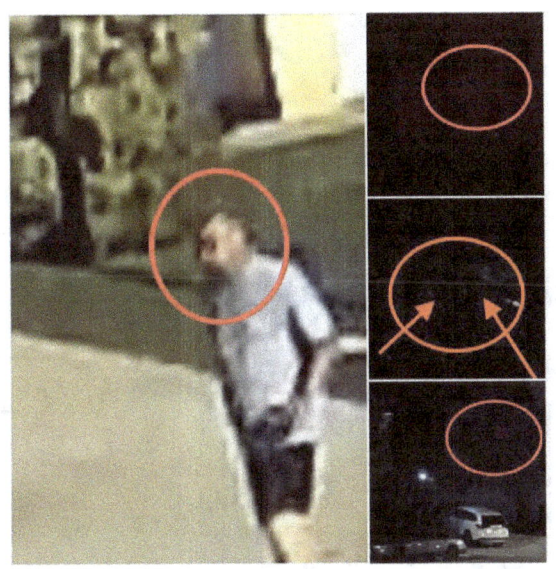

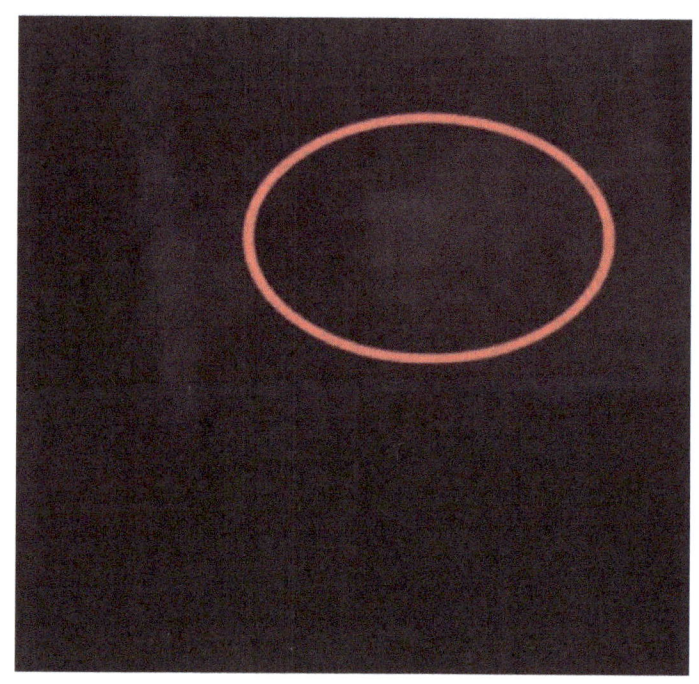

LAPD cops using Spear's Bedroom Parallel to me Upstairs with Sonic Weapon Setup, Leaving my Hands Trembling and Jerking uncontrollably the nest morning.

The image above, captured November 15, 2023, is the same house used by Black LAPD cops who continue with death threats using sagging pants so-called Gangsters and parolees to stalk me around town, and beamed assaults from here relentlessly. The homeowner, Spear is Black as well.

As shown, they had a Directed Energy Weapon blue antenna inside the homeowner's work truck that is empowered by the drone above they are using focused on me.

The following images reveal, top right, bottom left and bottom right, drone operated blue beam antennas inside trucks of neighbors. So, you are not confused, all three of these neighbors have white trucks.

Example Of Commercial, Green, Blue & Violet And Red Laser Beams

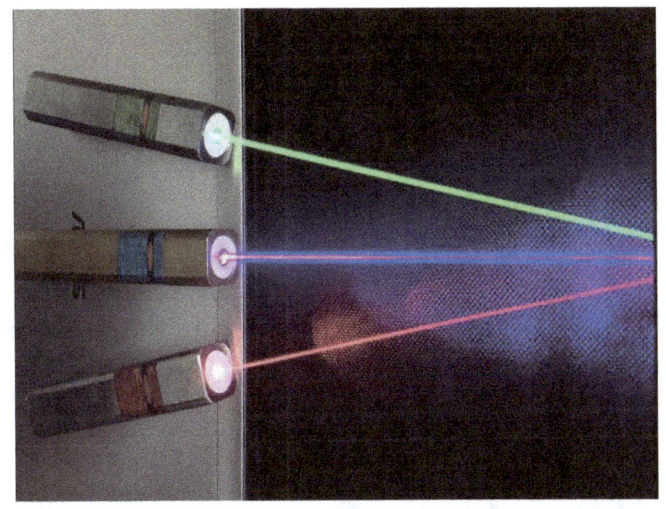

The Military Grade Weapon Beams are Red, Violet & Green, Etc., Wikipedia Blue Laser

A **blue laser** emits electromagnetic spectrum radiation with a wavelength between 400 and 500 nanometers, which the human eye sees in the visible spectrum as blue or violet

Blue lasers can be produced by:

- direct, inorganic diode semiconductor lasers based on quantum wells of gallium(III) nitride at 380-417nm or indium gallium nitride at 450 nm
- diode-pumped solid-state infrared lasers with frequency-doubling to 405nm
- up conversion of direct diode semiconductor lasers via thulium- or praseodymium-doped fibers at 480 nm
- metal vapor, ionized gas lasers of helium-cadmium at 442 nm and 10–200 maw
- argon-ion lasers at 458 and 488 nm

Lasers emitting wavelengths below 445 nm appear violet, but are nonetheless also called blue lasers. Violet light's 405 nm short wavelength, on the visible spectrum, causes fluorescence in some chemicals, like radiation in the ultraviolet (black light) spectrum (wavelengths less than 400 nm).

***

Los Angeles Resident's Rightful Lack of Trust In LAPD Using Military Weaponized Drones as cops use drones around me.

Los Angeles police pitched their plan to fly drones Wednesday night in a series of forums held across the city, the reaction was less positive than they might have hoped.

Dozens of residents expressed concern ranging from skepticism to outright opposition to any use of drones by the Los Angeles Police Department, telling department brass they were worried about their privacy, over-militarization of the police and that the devices would be flown far more than the LAPD has pledged.

The following images reveal the setup across the street where the black LAPD cops operate.

As shown, they have placed a Sonic Weapon in the front living room pointing in my direction attempting high-tech slow-kill murder.

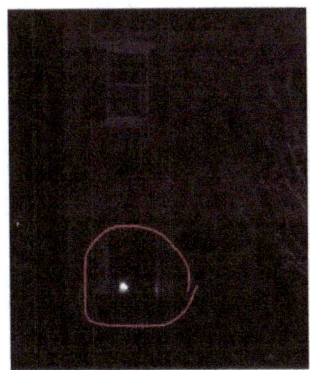 

DEADLY SONIC WEAPON FROM SPEAR'S HOUSE ACROSS THE STREET

Following are violet illuminated rooms in the homes USAF and Navy Personnel are using for beamed assaults

Bottom Image of Sonic Weapon System.

In the above image, also note the three circular lights seen which as stated previously is the main operation location for the military COINTELPRO, and Lockheed trainer. The three lights are the beamed Sonic Weapon setup.

The question is what is horrifically wrong with them?

A Personality Disorder Of Excessive Power Striving

Abstract

None of the existing formal diagnostic categories in psychiatry today addresses adequately the issues of excessive power-seeking, corruption and destructiveness. Excessive power strivings both poison the personality of the individual who is obsessed in his spirit and mind with power and do unacceptable harm to other peoples' lives. The present proposal of a diagnostic category of a Personality Disorder of Excessive Power Strivings is intended to fit into current diagnostic schema of DSM as well as into an earlier proposal (1) to examine in all psychopathology not only the burdens and damage people do and impose on their own selves and their own functioning, but also the harm they do to other peoples' lives and functioning. The diagnosis is to be used when the individual displays prolonged and severe manifestations of the following listed criteria: The basic feature which is always present in this personality disorder is:

1. Intense and extensive power strivings. In addition, at least three other of the following characteristics should be present;

2. Lack of empathy for people, and indifference to the suffering of others;

3. "Street smart" alertness and remarkable cunning committed to seizing and expanding power;

4. Ruthlessness in cultivation of power;

5. Scapegoating and projection of blame on to targeted individuals or a group, an insistent need to identify certain others as lowly, worthless and intended victims;

6. Corruption by power and addiction to power;

7. Demands of other people to be dependent on one's powerful personality, or that they become one's obedient followers;

8. Emphasis on symbolisms of pure vs. impure, holy vs. infidel, chosen vs. condemned;

9. A basic disrespect for the lives of others evidenced in callous or indifferent exposure of others to undue risks;

10. An absence of conscience in contexts of self-interest and opportunity;

11. A homicide/suicide orientation.

The technology is here to stay and the human mind is today's targeted battle field. Neuroweapons by beamed frequency manipulation can tamper with thoughts, creating depression, anger, fear to name a few capabilities.

It is difficult to believe that official personnel at the helm focus on influencing the destructions of marriages, families, friends, associates, business and personal relationships and children.

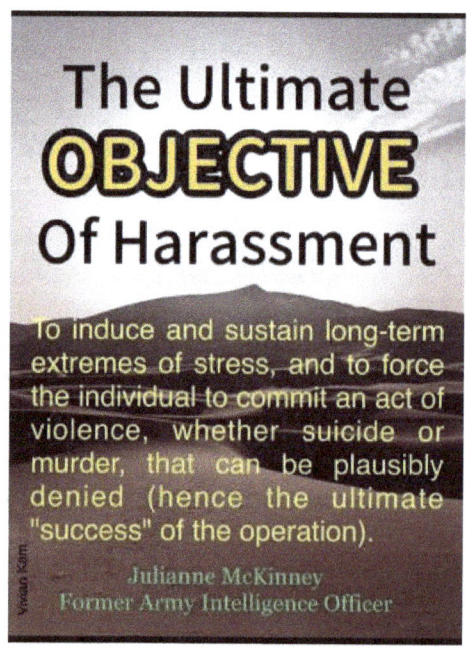

Following is another image taken of one of the four houses set up across the street.

Here again we see an image that I enlarged, of the upstairs setup. This is yet another example of what is satellite infused images in the background depicting so-called aliens. They know I am accurate about what I am reporting about the technology and how they are using it and continue hoping I would think aliens are setup around me.

The homeowner is seen in the center circled image with a red arrow pointing at him while sitting behind the computerized system with sunglasses on. This likely due to the glare of the monitor.

Below is a beam infused fake alien deception the third image down appearing to be an alien staring in my direction as I take the pic.

The background looks alien infused after enlargement with the bottom circled image clearly showing what again looks to me like a Gollum. I guess someone is a *Lord of the Rings* fan.

Frankly, again and again, this fraudulent program probably wish I could be convinced that this is an alien setup inside my neighbor's homes and alien technology this program is using, which some actually believing this. Believing this denies the brilliance of the human mind,

Nikola Tesla for example, who harnessed the energy of the Electromagnetic Spectrum for use for good understood for heinous use if in the wrong hands

Active Denial System "Pain Ray" Blue Laser used in Neighbor's vehicles for beamed assaults

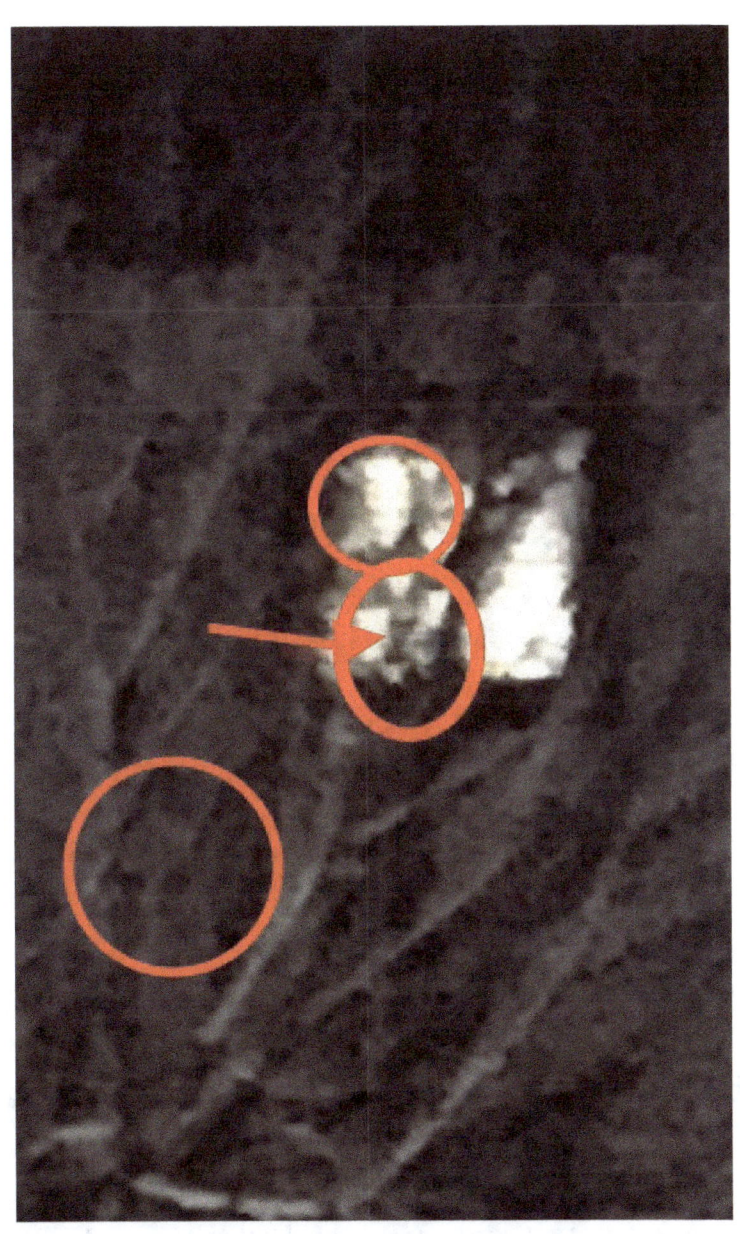

Note the Bottom Circled Image

As stated, previously, in 2008 they used a hologram to beam a similar image like this into my apartment. The problem here is that, again, I don't believe in aliens and am well versed on the technology used today.

Following is a closer view, of a neighbor's white truck, a better view ,parked in the front across the street belonging to the neighbor's home that federal agents are using today as their major operation field location. As soon as I posted the image of the antenna on several social media sites it was immediately taken out.

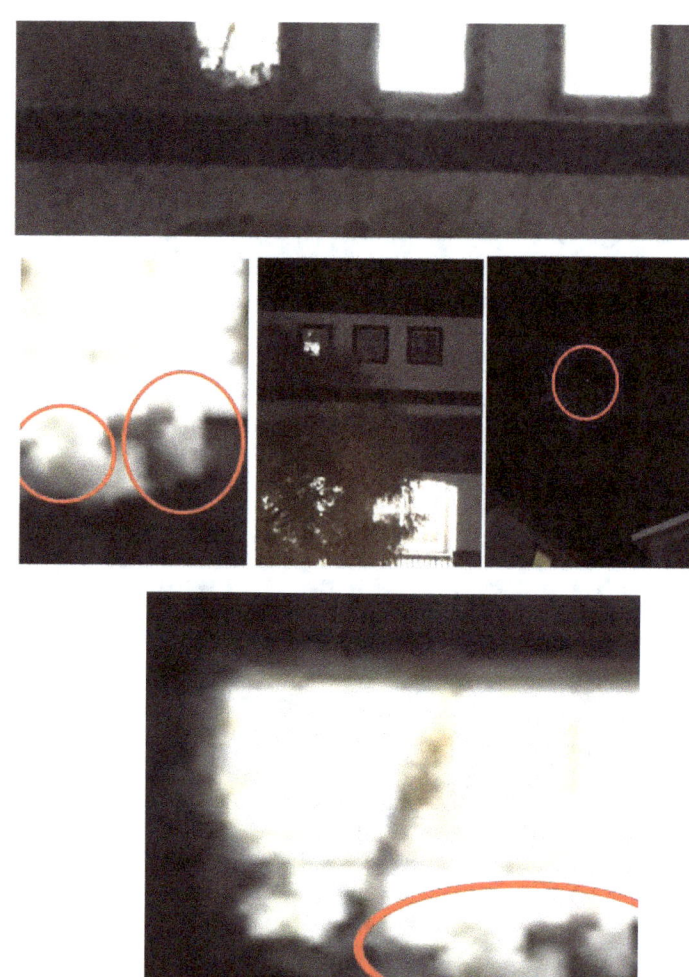

FBI Setup Across the Street

When I out the agents working the house, on the previous page, the home of the neighbor of one of the white truck with the blue beam antenna inside the truck, as shown in the following images, they moved next door to the house owned by the contracted Lockheed Black homeowner which again is also used by LAPD's Black heinous cops assigned to me.

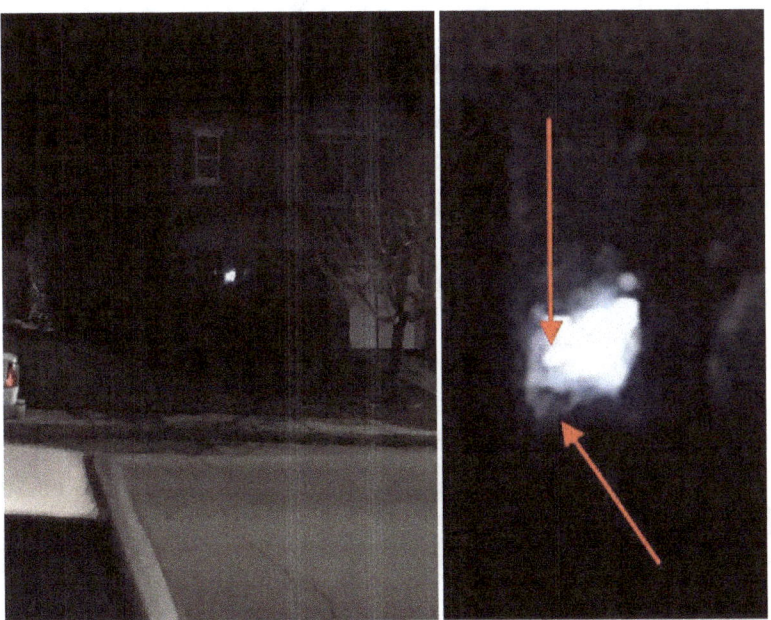

Above are images of the actual federal agents' setup across the street who have moved next door enlarged.

In my case they were blatantly wrong about me, and the story fed to the community. They operate close so when you leave home they also follow you around town at times making their presence known for intimidation.

Because they have to stick with the narrative told the community to motivate and use them, they use typical programmed racial stereotypes, today nearly 20 years later they are 100% still telling the public the same lies which simply are not true.

This is a program where no one is exempt from psychotronic toying with and while monitored, puppet people are influenced into playing specific roles and it all financed, again, as part of human experimentation from several budgets including the DOD covert operations Black Budget

A **black budget** or **covert appropriation** is a government budget that is allocated for classified or other secret operations of a state. The black budget is an account expenses and spending related to military research and covert operations. The black budget is mostly classified because of security reasons.

A black budget can be complicated to calculate, but in the United States it has been estimated to be over US$50 billion a year, taking up approximately 7 percent of the US$700 billion military budget.

***

A call went out to the United Nations across the USA and globally for recognition and hopeful full disclosure of mass human experimentation. Following is an excerpt entitled,

Inputs for psychosocial dynamics conducive to torture and ill-treatment report and a plea for relief:

This statement is with respect to the inputs requested for psychosocial dynamics conducive to torture and ill-treatment report (To inform the Special Rapporteur's annual interim report to be presented to the General Assembly at its 75th Session in October 2020)

This is an URGENT public interest petition to STOP the illegal and unauthorized abuses of advanced military-grade weapons that are being used for Torture Programs. Torture comprises of Mind-Reading, Mind control, Central Nervous System control, 24/7 anywhere tracking, Organized Gang-Stalking, 'Voices-To-Skull'('V2K'),

Physical Injury/harassment through Directed Energy Weapons. This has been going on in India for past 15 years at least (I am getting attacked/tortured for many years now, Voice to skull started in 2016)

All these attacks are 'no-touch' / 'Covert 'and **are** remotely operated - and so leave the minimum evidence (if at all) thus making all available laws ineffective and powerless to help the targeted innocent civilians. The people targeted and subsequently tortured systematically are termed as 'Targeted Individuals' (or 'TI'). There are several hundreds of 'TIs' defending and fighting for justice in India and globally across many countries now.

Total Surveillance, Mind-Reading, Body-Mind Control, Dream Manipulation - Using Neuroweapons - Remote Neural Monitoring Module ('Rnm') - Using this, harassers can view ALL the innermost thoughts of the targeted person on a screen - as clearly as one reads a newspaper. The eyes of the target become a live camera for the trackers. Whatever the targets view is recorded on the trackers' computer or viewed by the trackers' brains using a brain-to-computer interface

(BCI) / Brain-to-Brain interface (BBI)! These satellite-based technologies result in gross by- passing of fundamental human rights such as personal privacy, health, safety, data security, family security, etc. Pre-packaged dream sequences are routinely downloaded to TIs' brains and harassers interact with the victims while they are dreaming. Stressful traumas/shocks are also induced via artificial dreams (completely wirelessly - without any chip implants, electrodes etc.)...

Again, Lockheed Martin is heavily involved in Directed Energy Weapon Development and a key player around me with a huge facility less than 10 minutes away.

Lockheed's stated goals regarding Directed Energy development is revealed below:

Harnessing the Power of Lasers for 21st Century Security...

At sea, in the air and on the ground, Lockheed Martin is developing laser weapon systems ready to defend U.S. and allied forces. Combined with our platform integration expertise, these systems are designed to defeat a growing range of threats to military forces and infrastructure across all domains.

***

The following image taken of the corner house behind, when I again woke under painful beamed attacks. As with this image, when enlarged had similar images around it. However, seen is one of military personnel I see often in the main house across the street in front. He is now set up sitting at the window of the top far right bedroom in the corner house behind at angle to nearly parallel to where my bedroom is inside my house. The bright light in the background is the Sonic Weapon.

Military person image facing my bedroom upstairs that I capture when I woke under a full beamed assault. He is shown sitting in the top right corner room of the corner house behind.

Lockheed Top Secret Conglomerate On The Mission

SKUNK WORKS®

Lockheed Martin Skunk Works Minutes Away

At Lockheed Martin Skunk Works®, your mission defines our purpose.

Our team of dedicated engineers and scientists assume it can be done. With a visionary focus on the future, we partner with our customers to anticipate tomorrow's capability gaps and technology needs to solve the most critical national security challenges today.

With our enduring legacy, unique culture and way of operating, Skunks move quickly to develop disruptive solutions in core capability areas needed for our nation's future success. Discover how our team is defining the future by clicking the capability icons below.

Again, Lockheed is heavily involved and playing a major role in training on systems and devices nationwide and globally. This was confirmed, in my case, when I was out running errands and a craft followed me around town and as I headed home I saw it land on the Lockheed Martin rooftop

Nationwide suffering through experimentation by Directed Energy Weapons and monstrous covert torture continues.

The video details this official weapon system.

***

USAF Air Force Harnessing the Power of Directed Energy

Video Link

https://www.military.com/video/directed-energy-weapons

Realistically, I have come to accept that they keep tabs on any and every one exposing this program and by assignment cannot leave because of this. The goal is to shut you up again, by any means necessary.

This especially truth with my publications, books, blogs and website revealing accurately this program.

This also means that I have to continue exposing what is happening around me, with this program leaving me no choice if, not end up high-tech destroyed.

> You never owe your abusers anonymity. If the truth of their actions destroys their career, reputation, or relationships, that was their doing entirely.
>
> @deconstruct_everything

While they were crippling me previously, documented in *Covert Technological Murder – Pain Ray Beam* Book III, a military weaponized drone followed me around town in real-time and the operator using the beamed communication system laughed saying, *Hahaha… she can't walk.*

This was as they watched their handiwork as I limped barely able to make it to the door destroying both hip joints strategically. Highly credible exposure is a huge threat because the official overseeing what is happening today do not want what is written to become an obvious fact and mobilized the country against this type of criminal, hidden conduct.

Sadly, there is no stopping Government Issued (GIs), who are basic order followers dispatched into our communities who are brainwashed, indoctrinated and programmed after intense military behavioral modification.

Let's face it, militaries become official killing machines for governments all over the world. And again, these soldiers and cops are programmed, as part of employment, so intensely that it is pointless for them to accept the truth of decades of human experimentation and setups of thousands and this is an open literature, proven and historic fact.

Tortured When Good Soldiers Do Bad Things…

[Excerpt]

Normally, when the word *Torture* is written about in the media, we expect it to be either a historical account of a past event from decades ago like the Holocaust in World War II or to refer to human rights abuses in Asia, the Middle East, or the illegal use of torture during war conflicts and capturing terrorists, but always somewhere very far away from where we reside.

Reference Link

https://nhscorrupt.medium.com/tortured-when-good-soldiers-do-bad-things-versus-murder-for-profit-club-coined-by-nsa-whistle-185ce3749bd9

***

USAF House Different Upstairs Room Setup

Coincidentally, a Google search of the name of this homeowner for any military connection did reveal two actual people with the same name, a Staff Sergeant and an Officer in the USAF. The Officer is documented to be involved in technology. Admittedly, it could be a coincidence that the house across the street is owned by him. One thing for sure, military personnel are using one of the rooms and two specific individuals around me since 2010 and both seen in USAF uniforms and another in Navy uniform going and coming.

Military Weapons Drone and Car Antenna

Electromagnetic weapons are a type of directed energy weapons which use electromagnetic radiation to deliver heat, mechanical, or electrical energy to a target to cause pain or permanent damage. They can be used against humans, electronic equipment, and military targets generally, depending on the technology.

When used against equipment, directed electromagnetic energy weapons can operate similarly to omnidirectional electromagnetic pulse (EMP) devices, by inducing destructive voltage within electronic wiring. The difference is that they are directional and can be focused on a specific target using a parabolic reflector. Faraday cages may be used to provide protection from most directed and undirected EMP effects.

Necrosis is the death of body tissue. It occurs when too little blood flows to the tissue. When microwave beamed weapons repeatedly beam cook away body fluids, water and blood needed to sustain health the result is necrosis. Necrosis can be from injury, radiation, or chemicals. Necrosis cannot be reversed. When large areas of tissue die due to a lack of blood supply, another condition can also happen that is called gangrene.

True, Lockheed Martin doesn't actually run the U.S. government, but sometimes it seems as if it might as well. After all, it received $36 billion in government contracts in 2008 alone, more than any company in history. It now does work for more than two dozen government agencies from the Department of Defense and the

Department of Energy to the Department of Agriculture and the Environmental Protection Agency. It's involved in surveillance and information processing for the CIA, the FBI, the Internal Revenue Service (IRS), the National Security Agency (NSA), the Pentagon, the Census Bureau, and the Postal Service.

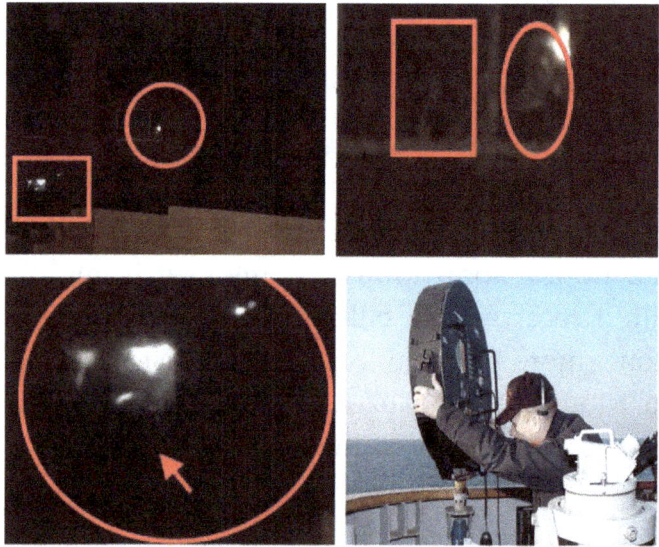

Corner House behind Sonic Weapon Setup

LAPD drone guiding an antenna placed in house directly behind me. As stated above three houses behind are being use. They are the corner house, massively, and the two next door to it.

Renee Pittman Website: The Cover-Up

Detailed Link

bigbrotherwatchingus.com

SILENCE IS NOT AN OPTION!

Approval for DoD Human Experimentation on Civilians

Procedures Governing the Activities of DOD Intelligence Components That Affect United States Persons:

Domestic Police Use of Military Weaponry

Note: Further details of the legalization of technology for civilian human experimentation can be found on the website associated with this book:

(Federal Efforts To Supply Surplus Military Weapons And Equipment To Domestic Law Enforcement Agencies) "The militarization of American policing has occurred as a direct result of federal programs that use equipment transfers and funding to encourage aggressive enforcement of the War on Drugs by state and local police agencies. One such program is the 1033 Program, launched in the 1990s during the heyday of the War on Drugs, which authorizes the U.S. Department of Defense to transfer military equipment to local law enforcement agencies.

[Excerpt]

12. This program, originally enacted as part of the 1989 National Defense Authorization Act, initially authorized the transfer of equipment that was 'suitable for use by such agencies in counter-drug activities.'

13. In 1996, Congress made the program permanent and expanded the program's scope to require that preference be given to transfers made for the purpose of 'counterdrug and counterterrorism activities.'

14. There are few limitations or requirements imposed on agencies that participate in the 1033 Program.

15. In addition, equipment transferred under the 1033 Program is free to receiving agencies, though they are required to pay for transport and maintenance. The federal government requires agencies that receive 1033 equipment to use it within one year of receipt,

16. so there can be no doubt that participation in this program creates an incentive for law enforcement agencies to use military equipment.

Source:

War Comes Home At America's Expense: The Excessive Militarization of American Policing, American Civil Liberties Union (New York, NY: ACLU, June 2014), p. 16.

WEAPONIZATION of GOVERNMENT

Congress funds the Federal Government including the FBI Fusion Centers. It is their duty to insure integrity, efficiency and Constitutionality.

- It is the duty of Federal Agencies to follow the rule of law and the Constitution, to be worthy of funding.
- Government's primary purpose is to see to the well-being and protection of the citizenry equally.
- Government's purpose is not to pervert their activity into harmful, biased abuse or targeting of "inconvenient" people like truth-seekers, truth-tellers, whistleblowers, peaceful activists, incorruptible or innocent citizens just to please corrupted bureaucrats or their cronies in the private sector.
- Government does not get to use separate and different criteria to base a person's rights upon. The Constitution guarantees equality.
- Government does not have the right to induce private citizens into covert vigilante criminality that ignores actual reality and bypasses due process, with secret accusations, secret evidence, secret witnesses, or, to circumvent the restrictions on government meant to rein in such abuses by anyone.
- Government does not have the right to embezzle taxpayer money to bribe or fund a secret standing citizen vigilante army to terrorize, harm, or kill any citizen for any reason, much less an innocent one.

This is why **FBI FUSION CENTERS** must be <u>defunded</u>, <u>closed</u>, <u>audited</u> to reveal their full records both official and unofficial, their actual, vice their claimed patterns and practices, their methods, their expenses, their tactics, equipment, their manipulation of the public, the nature of their claimed surveillance tech vice their weapons tech. Their "partners"(in crime) and their personnel as well as their paid citizen assassins, must be arrested, held indefinitely as domestic terrorists (per NDAA) *until* their fates are decided by military tribunal. Then a detailed investigation done as to the harm they caused exactly whom, based on what premise or excuse, upon whose orders, for how long, and the exact nature of that harm and its full repercussions upon that victim's life.

Ongoing Massive Directed Energy Weapon Attacks

Open Letter provided by NSA Analyst, and NSA Veteran, Whistleblower, Karen Melton-Stewart

This highly credible, vindictively and maliciously woman details the official, unified, high-tech targeting program, on her mission for full disclosure of this horrific crime. The above posters were created by her.

Many across the U.S.A. are working tirelessly working to expose this official MONSTROSITY as shown below.

***

Official Letter to Senator Warner, Chief of Senate Select Committee on Intelligence and Vice Chief Senator Marco Rubio

May 5, 2021

Senator Mark Warner
Chair Senate Intelligence Committee
703 Hart Senate Office Bldg.
Washington DC, 20510

Senator Marco Rubio
Vice-Chair Senate Intelligence Committee
284 Russell Senate Office Bldg.
Washington DC, 20510

Directed Energy Weapons (Dew) Being Used To Attack Civilians

Dear sirs,

Enclosed you will find two previously written letters in regard to the illegal use of Directed Energy Weapons (Weapons of Mass Destruction) by domestic terrorists inside the United States for many years now. One mailed to the Congressional DEW Caucus in 2019 warning them that these weapons had been loosed covertly on the American public and the other – my affidavit sent out to various entities as to my own investigation into why and how and by whom I had been illegally targeted as a legitimate whistleblower with these torturous, cruel, and lethal, high tech stealth weapons.

I am a retired National Security Agency Intelligence Analyst. I was attacked and illegally fired for asking the NSA IG George Ellard to investigate why my 6- month series of Top-Secret reports leading up to Operation Iraqi Freedom had been credited to another woman, who contributed not one iota but had slept her way through the managers in the NSA Weapons and Space Directorate. My supervisor USAF Major Sonja McMullen in Baltimore and co-worker Gary Willinski of Baltimore have direct knowledge of the fact that my work and the intended double promotion it garnered, was stolen. Yet, I was attacked, defamed, stalked, viciously harassed, set up and illegally fired (2010) in whistleblower retaliation. NSA withdrew their Security goons and Fusion Center run civilian InfraGard stalkers when they decided that they had won the illegal battle to set me up to be fired.

I sued NSA and my lawsuit was accepted by the EEOC (2010). It sat eight years on the docket. In 2015 when my lawyer applied for a subpoena that Implicated NSA SES Flag Badger in a break-in to my home multiple times and tampered with my work computer, NSA coerced the judge to ignore the subpoena, then arranged for the stalking harassment to start anew in Florida where I had temporarily moved in late 2014. When the PsyOps of vicious 24/7 stalking by FDLE/Fusion Center operatives and civilians failed to intimidate me, the Naval Security Group headquartered on Ft. Meade, MD. was asked to tell the Naval Security Group Base Commander to loan out his people and Directed Energy Weapons to a special project between the NSA, FBI, FDLE/FUSION CENTER, the Leon County Sheriff's Department and civilian entities like InfraGard, according to Sheriff's Deputy Jeffrey Canon. That special project was to infiltrate the Greenwood Hills neighborhood north of Tallahassee, spread rumors that I was a dangerous pedophile and had to be secretly murdered by Directed Energy Weapons for the good of all. And that the neighbors who cooperated or participated would be generously rewarded with tax dollars laundered through a third party into gift cards.

My affidavit shows you I approached every logical authority to no avail. The Leon County Sheriff's Department (LCSD) and authorities above them up to and including the AG of Florida, Pam Bondi, preferred to think or claim to think the DEW were fantasy and do nothing despite proof of outrageous microwave/electromagnetic meter readings and other evidence as given them by an ex-NSA Intelligence Analyst.

I am sickened that many victims have been reporting this to most state and law enforcement authorities (and their Federal representatives) and have been ignored, mocked, blown off and even insulted. Some have even been thrown into mental health facilities for reporting 24/7 torture and injury by clearly unauthorized people (Domestic Terrorists) wielding such devices that meet the criteria to do similar or the very same harm as unique to the Havana Syndrome. It would appear Pandora's box has been opened by rogue government entities, rogue law enforcement, rogue contractors, rogue military, and even a network of civilian mercenary gangs ready to set upon any victim a Fusion Center or military entity targets, regardless of the Constitution.

These devices must be made illegal, confiscated forthwith by a certain date, and it be made clear that possession of such a weapon henceforth carries a swift, severe, and immediate and irrevocable, unappealable punishment – even death. Otherwise, civilization simply cannot survive the proliferation of silent, deadly weapons any monkey can use, creating a doomed, predator – prey dystopia.

Karen M. Stewart
NSA Intelligence Analyst, 28 yrs.
Kams56@me.com

The cold, hard truth is that MK Ultra never ended and our military, today is heavily involved in ongoing, massive military mind control programs for DECADES, joined with specific alphabet agencies, contractors, and police which has in turn cause many whistleblowers, myself included, in a fight for our lives, truth the weapon indeed!

Sadly, we have been expertly, again, expertly effortlessly, programmed and brainwashed that those involved in this official, hideous program, military and police are heroes, no matter what when this specific type has consistently proven they are not. Yes, there are heroes but they are not involved in this program. Just because you put on a uniform does not make you one.

Please publicize this letter far and wide!

Trained Gang Stalkers And Their Idiotic Antics...

As part of official mobilized Organized Community Stalking, also known as Gang Stalking, Citizen Volunteers, Neighborhood Watch, etc., are provide an app to track the target around town.

They are told horrific untruths to keep the volunteers motivated and also vilify and degrade the target. There are reports of the mobilizing agency also using look a likes with fake negative images of the target in the App. The public becomes an arm of the illegal targeting, torture and military subliminal influence targeting program. As stated earlier, this program also uses ongoing Gaslighting tactics.

Everyone sees them: signs welcoming you to a neighborhood with the warning that *All Suspicious Persons and Activities Reported to Law Enforcement.* In the 1960's, Neighborhood Watch groups proliferated in supposed response to increased burglaries. In fact, the groups appear to have been a direct response to increased residential integration. A brainchild of the National Sheriffs Association, Neighborhood Watch groups were touted as a way to increase community involvement in crime prevention by encouraging residents to patrol their own streets and act as the eyes and ears of the local police.[3] But too often, local residents have interpreted this as a chance to become vigilantes, in many cases acting purely on bias to raise false alarms and profile fellow community members, endangering the very people these groups are allegedly designed to protect. The groups have proliferated across the country even as they have been demonstrated to promote profiling and distract from actual public safety, there being little evidence that Neighborhood Watch programs reduce crime.

***

Military Mind Control Experimentation Is Real!

Electromagnetic Fields from Monitors (**U.S. Patent 6506148 B2)**

In broad terms, *mind or thought control* can be defined as the manipulation of human subjects being in a way that causes them to act without autonomy, being unable to think independently and be subjected to beliefs and affiliations that they wouldn't freely choose.

Many conspiracy theories regarding mind control have been circling for years and a conspiracy theory is just that until proven it's a real fact.

The most horrifying thing about the theory of mind control is that it's more real than you may think, and the US has patented its use in 2001!

The one thing you haven't been taught in school is the effects that frequencies have on you. These frequencies can come in various forms, as everything you see, hear and feel is a frequency in its essence.

Frequencies have many different uses. They can be used to heal, open up your mind OR manipulate it. This is in fact what the US Government has been working *on since 1945.*

The main objective of the operation was to tap into the knowledge that *the Nazi scientists* have gathered from exposing human beings to unimaginable things.

Working together with the Nazis, the Government started developing a mind control program that would grant them power over their unsuspecting targets.

Ever since, the scientists that have devilishly consented to this kind of inhumane work they have been developing more and more advanced systems of mind control through frequencies.

One such patent is US 5356368 A, issued in 1994, which is named *Method of and apparatus for inducing desired states of consciousness.*

For the record, thousand's Targeted in this program are taken through various stages of experimentation including hopeful drug addiction destruction designed for Targets to sabotage themselves. This happens nationwide after intentional drug infiltration in specific communities while military intel, monitoring for use of war type bioweapons of mass destruction use PATENTED technology as shown below. The ones they try and try to influence, whose minds are too strong, and even technology can't keep them addiction trapped then become both a challenge and threat to these Workers of Iniquity by ongoing efforts to stop exposure of the truth, once awakened, designed to keep this program mind control efforts hidden.

This device uses pre-recorded brain waves that force your mind to feel what you are told to feel. It should be noted that 20 years after the fact, federal agents, and the Sheriff were trying to have someone plant narcotics around me. Again, I say, thank God their failed. This was to substantiate the false narrative they told the community

However, their quest doesn't stop there, as the most gruesome patent published yet is US 6506148 B2.

This patent is named **Nervous System Manipulation By Electromagnetic Fields From Monitors** and (filed in 2001) it was published in 2003.

The patent abstract reads:

Manipulating the nervous system is no joke at all. This kind of manipulation can affect everything from breathing to memory and intelligence! Now, how much do you watch TV, sit in front of your computer, or in short, how much are you exposed to monitors?

This means that your mind can be manipulated through your TVs, computer monitors and really any kind of monitor that you may see!

You are being manipulated through words, sound and images – **they are using frequencies to mess your mind up** and use it against your free will!

This is not a conspiracy theory – not anymore. It's real and it's been patented by the US Government!

How To Break Free?

Spend less time in front of screens, and more time in nature.

Increase your awareness about frequencies and work on techniques to raise and strengthen your own frequency.

Question everything, give everything a chance to consider, and do some research yourself before blindly believing in or dismissing claims and ideas.

I have personally stopped watching TV a long time ago, or to be exact, ever since I started working on my energetic frequencies. It seems that the more I raised my frequency, the more I saw TV as something very disturbing. Now, I don't even think about it.

Don't allow anyone influence you into doing things 'not by your own choice' or against your normal thinking process.

It's not your choice – it's their choice and you have been manipulated into thinking that you chose that!

Police Using Military Weaponized Drones

An officer's most critical value, aside from their heart and their compassion for people and dedication for service, is their integrity, he said. When that's lost, it jeopardizes everything they've touched.

With this technology in hand today, without a doubt integrity is lost with a belief of unaccountability and hidden high- tech manipulation.

***

Hundreds of criminal cases involving three city police officers charged earlier this month with falsifying evidence are now under review by prosecutors after corruption allegations sparked questions about whether their past police work could be suspect.

Prosecutors are already analyzing pending cases to determine if they can move forward on the strength of evidence other than the charged officers' testimony, but past cases and convictions — including those based on plea deals — could also be revisited, Los Angeles County Dist. Atty. Jackie Lacey said.

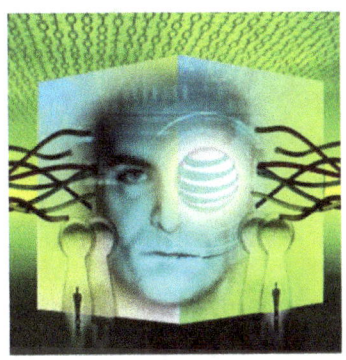

Remote Neural Monitoring

HAVE you ever thought about something you never shared with anyone, and have been horror-struck at the mere thought of someone coming to know about your little secret? If you have, then you probably have all the more reason to be paranoid now thanks to new and improved security systems being developed around the world to deal with terrorism that inadvertently end up impinging on one's privacy.

From personal experience, the Black cops in addition to beamed torture attacks from drones, also harass women using the patent, but also to terrorize both men and women sexually which equates to high-tech rape. This the result of drone aerial surveillance enabling voyeurism in home viewing technology leaving no privacy to shower or dress.

Below is a patent, that is reported use by ALL of these men, twisted military included. These sick individuals apparently get their kicks using the patent to sexually stimulate women and terrorize men and equates, basically to high-tech rape! Some of the seedier ones will also try to subliminally influence target's into sexual relationship if you are stupid so they can take advantage of the target in more ways than one! This is a powerful sexually stimulating patent demanding you try to relieve yourself. Sadly, many are watched in real-time thinking they are alone.

Beam Sexual Stimulation Patent

Pulsative manipulation of nervous systems

Patent number: 6091994

Abstract:

Method and apparatus for manipulating the nervous system by imparting subliminal pulsative cooling to the subject's skin at a frequency that is suitable for the excitation of a sensory resonance. At present, two major sensory resonances are known, with frequencies near 1/2 Hz and 2.4 Hz. The 1/2 Hz sensory resonance causes relaxation, sleepiness, ptosis of the eyelids, a tonic smile, a knot in the stomach, or sexual excitement, depending on the precise frequency used. The 2.4 Hz resonance causes the slowing of certain cortical activities, and is characterized by a large increase of the time needed to silently count backward from 100 to 60, with the eyes closed. **The invention can be used by the general public for inducing relaxation, sleep, or sexual excitement,** and clinically for the control and perhaps a treatment of tremors, seizures, and autonomic system disorders such as panic attacks.

Official personnel working this program as stated alternate shifts. While the target is sleeping they are still used for human experimentation which they hope is productive once awake.

***

Dream manipulation, patented technology and nightmares are both reported as part of the experimentation practiced in the wee hours of the morning and widely reported.

Scientific Lucid Dream Experimentation

https://patents.google.com/patent/US5551879A/en

https://www.boldbusiness.com/human-achievement/lucid-dream-technologies/

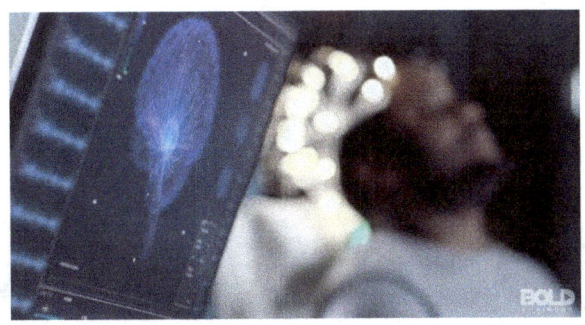

***

From *The Atlantic*: DARPA has dreamed for decades of merging human beings and machines. Some years ago, when the prospect of mind-controlled weapons became a public-relations liability for the agency, officials resorted to characteristic ingenuity. They recast the stated purpose of their neurotechnology research to focus ostensibly on the narrow goal of healing injury and curing illness. The work wasn't about weaponry or warfare, agency officials claimed. It was about therapy and health care. Who could object? But even if this claim were true, such changes would have extensive ethical, social, and metaphysical implications. Within decades, neurotechnology could cause social disruption on a scale that would make smartphones and the internet look like gentle ripples on the pond of history. [...]

Paying to use neighboring location is a common age-old tactic for the FBI, with funding, as shown in the official link below. The program within communities are supplementing the income of several around me making this heinous enterprise financially rewarding.

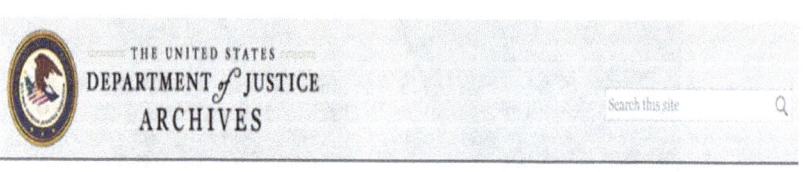

(For purpose of these Guidelines, an undercover operation involves fiscal circumstances if there is a reasonable expectation that the undercover operation will–

(a) Require the purchase or lease of property, equipment, buildings, or facilities; the alteration of buildings or facilities; a contract for construction or alteration of buildings or facilities; or prepayment of more than one month's rent;

NOTE: The purchase, rental, or lease of property using an assumed name or cover identity to facilitate a physical or technical surveillance is not an undercover operation for purposes of these Guidelines. However, since the expenditure of appropriated funds is involved, approval must be obtained from FBIHQ in conformance with applicable laws.

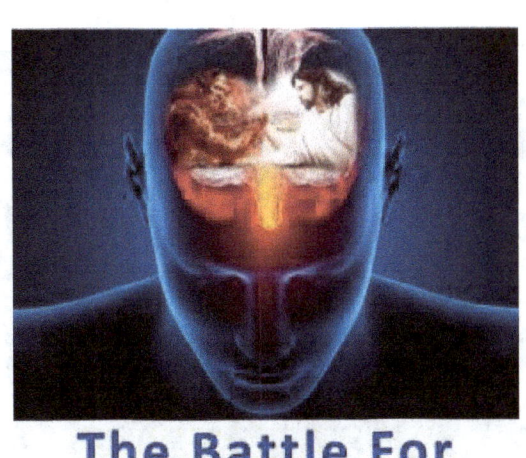

The Battle For Your Mind

With the U.S. Congress's approval of 30,000 weaponized drones included for US skies by 2020, with psychophysical beamed assault weapons and mind invasive Psychological Electronic (Psychotronic) beamed weapon systems the high-tech biometric targeting focus on anyone is taken to a whole new level.

The bottom image on the previous page is an actual image of one of three drones over my house 24/7 operated by official personnel setup around me.

***

While sitting comfortably before a weaponized operation system, in neighboring locations, or operation centers, a target can be destroyed in more ways than one, including murdered with no accountability. The human mind, body tissue, organs and joints, are deteriorated, now using microwave Directed Energy Weapons who are basically being slow cooked to death. Even while driving, I am hit with heart beam assaults.

Below is mobilized protest due to awareness of LAPD's historic corruption, who oppose drone use by LAPD and especially weaponized drones today given to Federal, state and local police departments. The FBI is also on record using drones since approximately 2006. The problem here is access to military weaponized drones.

NEARLY 1,500 US police departments operate drones but only about a dozen routinely dispatch them in response to 911 calls, according to this ACLU research link entitled (Atlas of Surveillance)

The bill requires the FAA to rush a plan to get as many drones in the air as possible within nine months

How many drones are we talking?

Shaun Waterman at The Washington Times reports the agency predicts that 30,000 drones could fill U.S. skies by the end of the decade.

***

Weaponized MQ-9 Reaper with camera GA

Naturally, many are concerned that surveillance by police and federal government agencies will skyrocket in response.

What is not publicized is the military drones are equipped with mind invasive technology psychotronics that demands legislature.

***

New human rights that would protect people from having their thoughts and other brain information stolen, abused or hacked have been proposed by researchers.

The move is a response to the rapid advances being made with technologies that read or alter brain activity and which many expect to bring enormous benefits to people's lives in the coming years.

Much of the technology has been developed for hospitals to diagnose or treat medical conditions, but some of the tools – such as brainwave monitoring devices that allow people to play video games with their minds, or brain stimulators that claim to boost mental performance – are finding their way into shops.

But these and other advances in neurotechnology raise fresh threats to privacy and personal freedom,

New arms threaten to destroy lives in strange ways. Directed energy weapons are among the high-tech arms of the century. They hurt and kill with electromagnetic power. Microwave weapons can be aimed at computers, electronical devices and persons. They have strong physical and psychological effects and can be used for military and terrorist activities. These weapons are also part of crimes (in Europe) that almost nobody knows except the victims and the offenders. Until now they make the perfect crime possible. No doubt, these weapons have a terrible future.

The best medicine for targets is to take the opportunity and redirect your focus into positive things that you love, with me gardening, and an online ecommerce store keep my mind occupied with positivity. Creating anything is great therapy, engrossing the mind and powerfully peaceful. In the mist of everything you can create beauty!

HYPOCRISY IS THE AUDACITY TO PREACH INTEGRITY FROM A DEN OF CORRUPTION

WES FESLER
PICTURE QUOTES .com

This information is excerpted from, Mind Control Technology - You Are Not My Big Brother Blog by Human Rights Advocate, Author Renee Pittman entitled, *The Blog THEY Don't Want You to see*

Each DEW can produce a range of effects from nonlethal to lethal, depending on factors such as the time on target, the distance to the target, and even the part of the target on which the DEW is focused. DEWs can use this range of effects to graduate responses to a threat. A graduated response could start with temporarily preventing use of an asset or its access to an area and increase to destruction of the asset if necessary (see fig. 2).

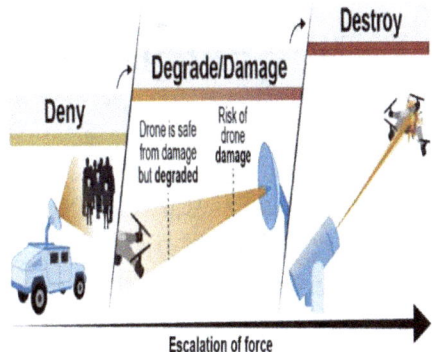

CHAPTER 8

Professional Gaslighting

When you stop living your life based on what others think of you real life begins. At that moment, you will finally see the door of self-acceptance opened. – Shannon L. Alder

> I was made and matured
> in the very mud
> my name was dragged through
> which is why I no longer idolize
> the words or works of people.
>
> I merely learn from them.
> —Morgan Richard Olivier

VIOLENCE AND GENDER Volume 3, Number 00, 2017 a Mary Ann Liebert, Inc. DOI:10.1089/vio.2017.0022

Review Article Mass Murder, Targeted Individuals, and Gang-Stalking: Exploring the Connection ~ Christine M. Sarteschi, PhD with an excerpt focused on me, Renee Pittman (Mitchell) and the death of Assistant District Attorney Myron May.

Abstract

People across the world refer to themselves as "targeted individuals" (TIs) and claim to be the victim of gang-stalking. The New York Times conservatively estimates that there are at least 10,000 people claiming to be victims of gang-stalking. Their perpetrators are typically perceived to be powerful government or law enforcement officials, who are seeking to destroy the life of the TI (Sheridan and James 2016). In retaliation, some have committed extreme violence. This article documents some of those cases and reviews the limited informational base of gang-stalking. These cases suggest that more research is needed to understand this unexplored belief system.

Keywords: gang-stalking, group-stalking, persecutory delusions, mass murder, attempted mass murder, targeted individual

Introduction

Individuals, worldwide, claim to be the target of organized group activities designed to cause them physical and psychological harm. They call themselves "targeted individuals" or "TIs." They refer to the organized group efforts to harm them as "gang-stalking." The New York Times estimates that there are at least 10,000 persons claiming to be the victims of gang-stalking (McPhate 2016). In retaliation for the perceived organized stalking, some have resorted to extreme violence in the form of mass murder or attempted mass murder.

[Excerpt]

Of notable interest is a lawsuit by **Renee Pittman Mitchell,** who corresponded with the Florida University shooter Myron May (a case examined later in this study) days before he shot several people in a campus library. Public court records from 2009 indicate that she had sought a restraining order against the Drug Enforcement Agency (DEA) "to stop the painful abuse she lives with daily" (Mitchell v. National Security Agency et al. 2009). She contended that the DEA

MASS MURDER, TARGETED INDIVIDUALS, AND GANG-STALKING 3

spoke to her, in the background of her cell phone calls, for the purpose of psychological torture. She also claimed that they used satellite technology to deliver harassing messages directly into her brain. The restraining order was denied, and the case was dismissed.

In 2015, she again sued the U.S. Government, this time alleging that her constitutional rights were being violated "by the United States' use of a Directed Energy Weapon to induce electronic harassment" against her (Mitchell v. United States 2016). She worried that they were conspiring with all levels of law enforcement and the military to commit a "convert technological murder" that she feared would be untraceable. Information available publicly revealed her to be a military veteran diagnosed with schizophrenia and a history of substance abuse.

Information contained within that lawsuit revealed that she also believed the following information: that the United States used energy weapons in her ceiling and in her neighbors homes for the purpose of "slowly diminishing" her joints and tissue causing "crippling organ damage;" that they have engaged in "mischievously tampering" of her publications by "inserting grammatical errors to discredit her as being inept and illiterate;" \that she is being monitored continuously; that the government conspired with her neighbors during her showers, and had left threatening messages in the background audio of her phone; that they have denigrated her by calling her a "whore;" that they have destroyed her relationship with her daughters; and they have prevented her from having personal relationships with anyone except her alleged attackers. She alleged that the purpose of the campaign against her was to prevent her from writing about the government's use of electromagnetic mind control technology. This had been the topic of her multiple self-published books. She also believed that the government was "exacting vengeance" for her having had received government benefit payments. Her 2015 case was **dismissed without prejudice.**

The Case of Myron May

According to police reports, 31-year old Myron DeShawn May returned to his alma mater, Florida State University (FSU). He entered the front lobby of the library at *12:30 a.m. on November 20, 2014. He was armed with a stolen .380 semiautomatic handgun. He then shot three people. One victim was an employee and alumnus of FSU, and the other two were students; all victims were male and his victim choice appeared to be random. After the shootings, Mr. May reloaded his weapon and exited the library, where he was confronted by police officers. When he refused to surrender, he was shot 24 times and died from his wounds.

In the aftermath of the shooting, it became evident that Mr. May had been experiencing mental health problems. The evidence that he left behind also suggests significant premeditation of the event. Students, on campus, observed him acting strangely for days before the shooting. One student said that he approached her and her classmates, identifying himself as an alumnus of the university and a lawyer (both of which were true). He claimed that Beyoncé and Jay-Z would be attending the university's homecoming and he needed their social media information to build support for the celebrities to visit campus. They described him as polite but pushy. Another student reported being present during a review session in class, days before shooting, when the instructor noticed an unfamiliar male student sitting in the class, who turned out to be Mr. May. When the instructor asked why Mr. May was there, he responded that he was a former student and was just sitting in for the day. No further action was taken.

Mr. May recorded multiple videos 2 days before the shooting. Each video was accompanied by a script. The first video was *41 min long and described some of his experiences as a TI. The goal of "gang-stalking," he said, was to drive the TI "crazy." Gangs-stalking programs, he went on to explain, involved "Control Panel Stalkers," whom he described as a group of people confined to an operation center whose job was to monitor TIs across America. He complained of being harassed by people who wore dark sunglasses and who would give him "strange looks." He heard voices that would narrate his movement through his apartment. Gang-stalkers also would shine bright lights into his home and vehicle, a tactic he called "brightening." They also subjected him to a "noise campaign" which prevented him from sleeping and they would break into his apartment to rearrange his household items, all tactics designed to cause insanity.

Other methods of psychological torture included "mopping," a scenario that involved 5–10 cars driving past him, all at once, each driver wearing dark sunglasses; "hacking," a tactic that would involve people gaining remote access to his electronic devices and preventing his downloading of books about gang-stalking; electronic harassment in the form of "directed energy weapons"; and the usage of psychotronic weapons (also known in the gang-stalking community as voice of God technology) that utilized microwave technologies to induce sound into the ears of TIs and would allow the gang-stalkers to see images directly from his mind as he slept.

The second video left by Mr. May was playing on a loop on his computer when it was found by authorities. In this

MASS MURDER, TARGETED INDIVIDUALS, AND GANG-STALKING 5

video, which was *33min long, he prays, asks for forgiveness, and says goodbye to his family and friends. The third video, also about 33 min long, provides a deeper explanation of his experience as a TI and his rationale for murder. His main motivation was to focus media attention on the plight of TIs. The media attention would enable other TIs to "have a shot at living a normal life." The life that he was denied. He insisted that he was not mentally ill, but was instead a victim of targeted harassment.

Text messages revealed that he had been interacting with Rene Pittman Mitchell, the self-published author of multiple gang-stalking books (Remote Brain Targeting; Diary of an Angry Targeted Individual: Remote Neural Monitoring; Convert Technological Murder: Big Brother Approved! and The Targeting of Myron May: Florida State University Gunman, among others), who, as described above, twice sued the U.S. Government regarding her beliefs about being spied upon. Ms. Pittman was initially contacted by Mr. May through social media. As their interactions continued she became worried that he was an "imposter" and attempted to cut ties with him. It appears as though he continued to contact her despite her attempt to end their relationship. Before the shooting, he mailed her a certified package and left her multiple voice mails and text messages. He also sent her an email at 11:19 p.m., the night of the shooting that said "I've been getting hit with the direct energy weapon in my chest all evening. It hurts really bad now." The two also appeared to be having a disagreement about his threats of suicide. They exchanged the following messages, in between several phone calls to one another, beginning after 9 p.m., on the night of the shooting:

Pittman: "It appears you are being used to play on my humanity side with the suicide crap."

May: "No, Renee. Stop being paranoid. I'm a genuine guy. I'm still getting hit right now."

Pittman: "You never mailed anything and you know it. I guess the HNIC thought they had a handle on me through my sympathy. You are a shill! You say you want to kill yourself. Go for it. That is your business and not mine. Sorry kid we must part ways now. I can't save you."

May: "Calm down now!!! My death can't be in vain. I devised a scheme and you are about to mess it up real bad Renee." Pittman: "If you kill yourself you hurt our effort and allow this program to continue. You can bet the media will portray you as schizophrenic. saying you were hearing voices. Don't involve me."

Myron May also left a five-page explanation about being harassed by gang-stalkers titled, "My Experiences as a Targeted Individual." In it, he expressed regret for having resorted to a mass shooting, but felt that his options were extremely limited. By virtue of being a TI, he believed that everything had been taken away from him. "I have literally been robbed of life through psychological, financial, and emotional hardship." The letter details his experiences with electronic harassment. The letter explained: when it started, who was following him, and what he believed was the factual daily experience of a TI. He hoped that his actions would benefit other TI victims.

***

The fact is Myron May threw in the towel with the understanding that a life he had worked so hard for and excelled within his profession was over. An example of what appears to be a growing collusion of denial of these PATENTED weapon, those who have not actually looked into contrary publications, and weapons in research, testing and development programs for decades appears below. Discrediting the existence of again, patented weapons, systems and devices, can be seen in an article targeting Dr. John Hall, who has published two books on US citizens used as Guinea Pigs and satellite terrorism. In his case the connecting him is an even more bizarre narrative where this group of so-called experts, connect people reporting official stalked by Stasi Zersetzung tactics, outrageously associate the stalking by mobilized efforts, Military Reserve, Neighborhood Watch, and police mobilized Citizen Volunteers, connected to *Haunted People Syndrome.*

The Dr. John Hall story: *A Case Study in Putative Haunted People Syndrome*

Ciaran O'Keeffe, James Houran, Damien J. Houran, Neil Dagnall, Kenneth Drinkwater, Lorraine Sheridan & Brian Laythe Pages 910-929 | Received 28 Jun 2019, Accepted 23 Sep 2019, Published online: 17 Dec 2019

Abstract:

Recurrent and systematic perceptions of anomalous *subjective* and *objective* anomalies.

Such signs or symptoms are traditionally attributed to spirits and the supernatural, but these themes are hypothesized to morph to surveillance and stalking in reports of group-(or gang) stalking. We tested this premise with a quali-quantitative exercise that mapped group-stalking experiences from a published first-hand account to a Rasch measure of haunt-type anomalies. This comparison found significant agreement in the specific signs or symptoms of both phenomena. Meta-patterns likewise showed clear conceptual similarities between the phenomenology of haunts and group-stalking. Findings are consistent with the idea that both anomalous episodes involve the same, or similar, attentional or perceptual processes and thereby support the viability of the HP-S construct.

Lastly another example is an article written by

.Duke University Press: *The Technical Delusion* skirting the Ongoing Mind Control "Mass Delusions" brilliantly orchestrated, official coverup

Duke University Press: *The Technical Delusion*

***

The Technical Delusion: Electronics, Power, Insanity by Jeffrey Sconce, Publication date: 2019

Delusions of electronic persecution have been a preeminent symptom of psychosis for over two hundred years. In *The Technical Delusion* Jeffrey Sconce traces the history and continuing proliferation of this phenomenon from its origins in Enlightenment anatomy to our era of global interconnectivity. While psychiatrists have typically dismissed such delusions of electronic control as arbitrary or as mere reflections of modern life, Sconce demonstrates a more complex and interdependent history of electronics, power, and insanity.

Drawing on a wide array of psychological case studies, literature, court cases, and popular media, Sconce analyzes the material and social processes that have shaped historical delusions of electronic contamination, implantation, telepathy, surveillance, and immersion. From the age of telegraphy to contemporary digitality, the media emerged within such delusions to become the privileged site for imagining the merger of electronic and political power, serving as a paranoid conduit between the body and the body politic. Looking to the future, Sconce argues that this symptom will become increasingly difficult to isolate, especially as remote and often secretive powers work to further integrate bodies, electronics, and information. "

***

Duke University Brain-To-Brain Interface

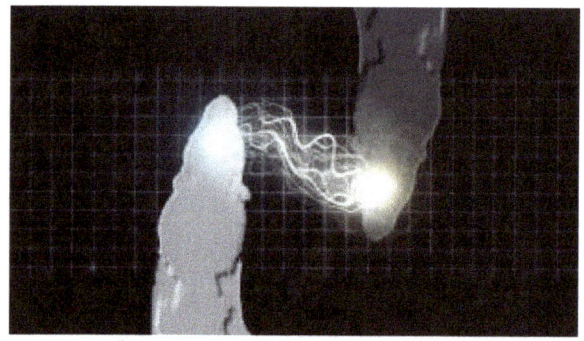

Duke University And Mind

Control Experimentation

***

Center For Cognitive Neuroscience

https://dibs.duke.edu/research/centers/ccn

***

The Pentagon's interest in the brain for everything from mind-controlled vehicles to advanced prosthetics.

"Whether or not the Duke University experiments turn out to be historic (some skepticism has already been raised), the work reflects a growing Pentagon interest in neuroscience for applications that range from such far-off ideas as teleoperation of military devices (think mind-controlled drones), to more near-term and less controversial technology, like prosthetics controlled by the human brain. In fact, like the Arpanet, the experiment on the rat "brain net" was sponsored by the Defense Advanced Research Projects Agency (DARPA)."

***

Duke Patents Mind Control Weapons

"Work on Brain-Machine Interface (think monkey controlling a joystick with its thoughts) is old news, but a patent granted earlier this month underscores researchers' confidence that a broader set of military applications is possible: like controlling weapons with your mind. In "Apparatus for acquiring and transmitting neural signals and related methods," researchers at Duke University [...]"

***

Duke Researchers Envision Mind Controlled Weaponry

The *Blue Devils* certainly haven't fared well on the hardwood of late, but a group of engineering minds at Duke University are thinking up ways to get even. While we can't actually confirm the motives, a recent patent filing spells out details of a device that can use the brain's thoughts to control an array of mechanical and electrical devices, up to and including weapons. Thought-controlled interfaces have long since been available, but these researchers are suggesting that everything from household items like televisions and ovens to weapons systems could be used to not only improve one's quality of life, but could actually produce thoughts that literally kill. Interestingly, the verbiage even mentions that the recently-ratified UWB technology could be used to beam commands from your devious brain, and although it's not surprising to find that DARPA has a hand in funding department, we're still not any closer to finding out when our military will switch from triggers to impulses

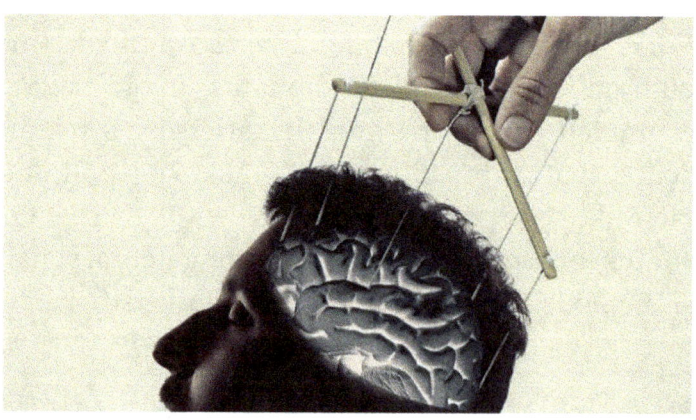

When Altering Brain Functions Becomes Mind Control

Abstract

Functional neurosurgery has seen a resurgence of interest in surgical treatments for psychiatric illness. Deep brain stimulation (DBS) technology is the preferred tool in the current wave of clinical experiments because it allows clinicians to directly alter the functions of targeted brain regions, in a reversible manner, with the intent of correcting diseases of the mind, such as depression, addiction, anorexia nervosa, dementia, and obsessive-compulsive disorder. These promising treatments raise a critical philosophical and humanitarian question. Under what conditions does altering brain function qualify as mind control? In order to answer this question, one needs a definition of mind control. To this end, we reviewed the relevant philosophical, ethical, and neurosurgical literature in order to create a set of criteria for what constitutes mind control in the context of DBS. We also outline clinical implications of these criteria. Finally, we demonstrate the relevance of the proposed criteria by focusing especially on serendipitous treatments involving DBS, i.e., cases in which an unintended therapeutic benefit occurred. These cases highlight the importance of gaining the consent of the subject for the new therapy in order to avoid committing an act of mind control.

Real Life Mind Control Governments Are Working On

Interesting Engineering reported on real-life mind-control technologies and mentioned that last year UB's Artificial Intelligence Institute received a grant to accelerate research into biometric information gathering from brain waves and eye movements — while playing a computer game.

The idea is to eventually scale up to 250 aerial and ground robots, working in highly complex situations, said Souma Chowdhury, assistant professor of mechanical and aerospace engineering in the School of Engineering and Applied Sciences.

***

The BRAIN Initiative: developing technology to catalyze neuroscience discovery

Abstract

The evolution of the field of neuroscience has been propelled by the advent of novel technological capabilities, and the pace at which these capabilities are being developed has accelerated dramatically in the past decade. Capitalizing on this momentum, the United States launched the Brain Research through Advancing Innovative Neurotechnologies (BRAIN) Initiative to develop and apply new tools and technologies for revolutionizing our understanding of the brain. In this article, we review the scientific vision for this initiative set forth by the National Institutes of Health and discuss its implications for the future of neuroscience research. Particular emphasis is given to its potential impact on the mapping and study of neural circuits, and how this knowledge will transform our understanding of the complexity of the human brain and its diverse array of behaviors, perceptions, thoughts and emotions.

For the benefit of many today, receiving funding for ongoing high-tech human experimentation, remote neural monitoring and remote viewing, Brain Computer Interface, and various other types of advancing technologies, derived from the Electromagnetic Spectrum, around the Earth and in use since the beginning of time, there really is no need to spell it out for them OF which they are fully aware.

MK Ultra wasn't just one project, as the US Supreme Court wrote in a 1985 decision on a related case. It was 162 different secret projects that were indirectly financed by the CIA, but were contracted out to various universities, research foundations and similar institutions." In all, at least 80 institutions and 185 researchers participated.

The initial studies with the USA used the University of Maryland and George Washington University for some of its top secret MKULTRA experiments in behavior control in the 1950s and 1960s. The Director of Central Intelligence, testified that the C.I.A. had secretly supported human behavior control research at 80 institutions, including **44 colleges or universities** as well as hospitals, prisons and pharmaceutical companies. Below is the short list.

Though available information on the full scope of MKULTRA, is limited, lists of the participating institutions still exist, and, according to those lists, Indiana University (listed as University of Indiana) was one of the universities that conducted research for the project among many.

The similarities between Stranger Things and US involvement in mind control projects are CHILLING

According to Wikipedia, "MKULTRA used numerous methodologies to manipulate people's mental states and alter brain functions, including the surreptitious administration of drugs (especially LSD) and other chemicals, hypnosis, sensory deprivation, isolation and verbal abuse, as well as other forms of psychological torture.

Louis Jolyon West And The UCLA Mind Control Program

Abstract:

Louis Jolyon (Jolly) West, M.D. (1924-1999) was a well-known Los Angles psychiatrist who served as the chair of UCLA's Department of Psychiatry and as director of the UCLA Neuropsychiatric Institute from 1969 to 1989. He was an expert on cults, coercive persuasion (brainwashing), alcoholism, drug abuse, violence, and terrorism. The collection contains Dr. West's research materials, lecture and presentation materials, personal and professional correspondence, and documents related to his professional associations and academic positions.

His specialty was LSD and was an expert hired by the CIA during MK Ultra LSD experimentation. He was also well known as an expert on behavioral modification, brainwashing, mind control, trauma-based mind control, mind and dissociative, such as Project Bluebird and various techniques used for shattering and dividing the mind, torture, substance abuse, post-traumatic stress disorder and use of violence.

Reference: *Finding Aid for the Louis Jolyon West Papers LSC.0590*

Synthetic/Artificial Telepathy" Beamed (High-Tech Schizophrenia) Patent and Details

https://patents.google.com/patent/WO2005055579A1/en

***

https://www.socsci.uci.edu/newsevents/news/2013/2013-05-10-artificial-telepathy-to-create-pentagon-s-telepathic-soldiers.php

Across the nation many report patented technology used as part of the 24/7 harassment program The patent below *Voice to Skull* (V2K) technology is just one of many patents. The DOD Voice of God is another example of technology also known as Synthetic/Artificial Telepathy by the USAF, Neural Decoding, Neurophone, Microwave Auditory Effect (Frey Effect), etc.

Patent For Microwave Voice-To-Skull Technology
5-9-3

Patent For Microwave Voice-To-Skull Technology – High-Tech Beamed Official Schizophrenia destroy the lives of thousands

United States Patent 4,877,027 Brunkan October 31, 1989

Hearing System

Abstract

Sound is induced in the head of a person by radiating the head with microwaves in the range of 100 megahertz to 10,000 megahertz that are modulated with a particular waveform. The waveform consists of frequency modulated bursts. Each burst is made up of ten to twenty uniformly spaced pulses grouped tightly together. The burst width is between 500 nanoseconds and 100 microseconds. The pulse width is in the range of 10 nanoseconds to 1 microsecond. The bursts are frequency modulated by the audio input to create the sensation of hearing in the person whose head is irradiated.

Inventors: Brunkan; Wayne B. (P.O. Box 2411, Goleta, CA 93118) Appl. No.: 202679 Filed: June 6, 1988

Current U.S. Class: 607/56 International Class: A61N 005/00 Field of Search: 128/420.5,804,419 R,421,422,746 381/68

Other References

Cain et al, *Mammalian Auditory Responses* . . . , IEEE Trans Biomed Eng, pp. 288-293, 1978.

Frey et al, *Human Perception . . . Energy Science,* 181,356-358, 1973.

Jaski, *Radio Waves & Life,* Radio-Electronics, pp. 45-45, Sep. 1960.

Microwave Auditory Effects and Applications, Lin, 1978, pp. 176-177.

Primary Examiner: Cohen; Lee S.

Attorney, Agent or Firm: Brelsford; Harry W.

Claims

I claim:

1. Apparatus for creating human hearing comprising:

a. an audio source for creating electrical audio waves having positive peaks;

b. a frequency modulator generator connected to the audio source to create frequency modulated bursts;

c. a source of constant voltage to create a voltage standard that is in the range of 25% to 85% of the peak voltage of the audio waves;

d. a comparator connected to the voltage source and the audio source to compare the instantaneous voltage of the waves from the audio source with the voltage standard;

e. a connection of the comparator to the frequency modulator generator to activate the frequency modulator generator when the instantaneous voltage of the audio wave exceeds the standard voltage;

f. a microwave generator creating microwaves in the range of 100 megahertz to 10,000 megahertz and connected to the frequency modulator generator, generating microwaves only when pulsed by the frequency modulator generator; and

g. an antenna connected to the microwave generator to radiate the head of a human being to produce the sounds of the audio source.

2. Apparatus as set forth in claim 1 wherein the frequency generating range of the frequency modulator generator is 1 Khz to 100 KHz for bursts and 100 KHz to 20 MHZ for pulses within a burst.

3. Apparatus as set forth in claim 1 wherein the frequency generating range of the frequency modulator generator is one Khz to 100 KHz for bursts and 100 KHz to 20 MHZ for pulses within a burst and the duration of each pulse of the frequency modulator generator is in the range of 10 nanoseconds to 1 microsecond.

4. Apparatus as set forth in claim 1 wherein the voltage standard is approximately 50% of the peak of the audio waves.

5. Apparatus as set forth in claim 1 wherein the antenna is of the type that projects the microwaves in space to the head of a person.

6. Apparatus for creating human hearing comprising:

   a. an oscillator creating an electromagnetic carrier wave at a selected frequency in the range of 100 Mhz to 10,000 Mhz;

   b. a pulse generator connected to said oscillator to pulse the carrier with pulses having a width in the range of 10 nanoseconds to 1 microsecond with a minimum spacing between pulses of about 25 nanoseconds;

   c. a frequency modulator connected to the pulse generator;

   d. an audio signal generator connected to the modulator which modulates the pulses in accordance with the audio signal; and

   e. a transmitting antenna connected to the oscillator to transmit the carrier wave as thus modified to project the electromagnetic energy through space to the head of a person.

7. Apparatus as set forth in claim 6 wherein the modulator is a frequency modulator to vary the density of bursts within an audio envelope as a function of the audio amplitude.

8. The method of irradiating a person's head to produce sound in the head of the person comprising

   a. irradiating the head of a person with microwaves in the range of 100 Mhz to 10,000 Mhz;

   b. pulsing said microwaves with pulses in the range of 10 nanoseconds to 1 microsecond; and

   c. frequency modulating groups of pulses called bursts by audio waves wherein the modulation extends from 1 Khz to 100 Khz.

Description

This invention relates to a hearing system for human beings in which high frequency electromagnetic energy is projected through the air to the head of a human being and the electromagnetic energy is modulated to create signals that can be discerned by the human being regardless of the hearing ability of the person.

The Prior Art

Various types of apparatus and modes of application have been proposed and tried to inject intelligible sounds into the heads of human beings. Some of these have been devised to simulate speech and other sounds in deaf persons and other systems have been used to inject intelligible signals in persons of good hearing, but bypassing the normal human hearing organs.

U.S. Pat. No. 3,629,521 issued Dec. 21, 1971 describes the use of a pair of electrodes applied to a person's head to inject speech into the head of a deaf person. An oscillator creates a carrier in the range of 18 to 36 KHz that is amplitude modulated by a microphone.

Science magazine volume 181, page 356 describes a hearing system utilizing a radio frequency carrier of 1.245 GHz delivered through the air by means of a waveguide and horn antenna. The carrier was pulsed at the rate of 50 pulses per second. The human test subject reported a buzzing sound and the intensity varied with the peak power.

Similar methods of creating clicks inside the human head are reported in I.E.E.E. Transactions of Biomedical Engineering, volume BME 25, No. 3, May 1978.

The transmission of intelligible speech by audio modulated Microwave is described in the book Microwave Auditory Effects and Applications by James C. Lin 1978 publisher Charles C. Thomas.

Brief Summary Of The Invention

I have discovered that a pulsed signal on a radio frequency carrier of about 1,000-megahertz (1000 MHz) is effective in creating intelligible signals inside the head of a person if this electromagnetic (EM) energy is projected through the air to the head of the person. Intelligible signals are applied to the carrier by microphone or other audio source and I cause the bursts to be frequency modulated. The bursts are composed of a group of pulses. The pulses are carefully selected for peak strength and pulse width. Various objects, advantages and features of the invention will be apparent in the specification and claims.

Brief Description Of The Drawings

In the drawings forming an integral part of this specification:

FIG. 1 is a block diagram of the system of the invention.

FIG. 2 is a diagram of an audio wave which is the input to be perceived by the recipient.

FIG. 3 is a diagram on the same time coordinate as FIG. 2 showing bursts that are frequency modulated by the wave form of FIG. 2.

FIG. 4 shows, on an enlarged time coordinate, that each vertical line depicted in FIG. 3 is a burst of pulses. (A burst is a group of pulses).

FIG. 5 shows, on a further enlarged time coordinate, a single continues pulse, Depicted as a vertical line in FIG. 4.

Detailed Description of the Invention

Inasmuch as microwaves can damage human tissue, any projected energy must be carefully regulated to stay within safe limits. The guideline for 1,000 MHz, set by the American Standards Institute, is 3.3 mw/cm2 (3.3 milliwatts per square centimeter). The apparatus described herein must be regulated to stay within this upper limit.

Referring to FIG. 1 a microphone 10 or other generator of audio frequencies, delivers its output by wire 11 to an FM capable pulse generator 12 and by branch wire 13 to a comparator 14. The comparator 14 also receives a signal from a voltage standard 16. When the peak voltage of the audio generator 10 falls below the standard 16 the comparator delivers a signal by wire 17 to the FM capable pulse generator 12 to shut down the pulse generator 12. This avoids spurious signals being generated. The output of the FM pulse generator 12 is delivered by wire 18 to a microwave generator 19 which delivers its output to the head of a human being 23. In this fashion the person 23 is radiated with microwaves that are in short bursts.

The microwave generator 19 operates at a steady frequency presently preferred at 1,000 megahertz (1,000 million). I presently prefer to pulse the microwave energy at pulse widths of 10 nanoseconds to 1 microsecond. For any one setting of the FM capable generator 12, this width is fixed. The pulses are arranged in bursts. The timing between bursts is controlled by the height of the audio envelope above the voltage standard line. In addition, the bursts are spaced from one another at a non-uniform rate of 1 to 100 KHz. This non-uniform spacing of bursts is created in the FM capable generator 12.

Referring to FIG. 2 there is illustrated an audio wave 27 generated by the audio input 10 wherein the horizontal axis is time and the vertical axis is voltage. For illustrative purposes the wave 27 is shown as having a voltage peak 28 on the left part of FIG. 2 and a voltage peak 29 of the right side of FIG. 2. The voltage standard 16 of FIG. 1 generates a dc voltage designated at 31 in FIG. 2. This standard voltage is preferably at about 50% of the peak voltage 28. The comparator 14 of FIG. 1 actuates the FM capable generator 12 only when the positive envelope of the audio wave 27 exceeds the voltage standard. The negative portions of the audio wave are not utilized.

Referring now to FIG. 3 there is illustrated two groups of bursts of microwave energy that are delivered by the antenna 22 of FIG. 1 to the head of the person 23. FIG. 3 has a horizontal time axis identical to the time axis of FIG. 2 and has a vertical axis that in this case represents the power of the microwaves from generator 19. At the left part of FIG. 3 are a plurality of microwave bursts 32 that occur on the time axis from the point of intersection of the standard voltage 31 with the positive part of the audio wave 27, designated as the time point 33 to time point 34 on FIG. 2. It will be noted in FIG. 3 that the bursts 32 are non-uniform in spacing and that they are closer together at the time of maximum audio voltage 28 and are more spread out toward the time points 33 and 34. This is the frequency modulation effected by the FM pulse generator 12.

Referring to the right part of FIG. 3 there are a plurality of microwave bursts 36 that are fewer in number and over a shorter time period than the pulses 32. These extend on the time axis of FIG. 2 from point 37 to point 38. These bursts 36 are also frequency modulated with the closest groupings appearing opposite peak 29 of FIG. 2 and greater spacing near time points 37 and 38.

Referring now to FIG. 4 there is illustrated the fact that a single burst shown as straight lines 32 or 36 on FIG. 3 are made up of ten to twenty separate microwave pulses. The duration of the burst is between 500 nanoseconds and 100 microseconds, with an optimum of 2 microseconds. The duration of each pulse within the burst is 10 nanoseconds to 1 microsecond and a time duration of 100 nanoseconds is preferred. The bursts 32 of FIG. 3 are spaced non-uniformly from each other caused by the frequency modulation of 12. FIG. 4 depicts a burst. Each vertical line 40 in FIG. 4 represents a single pulse. Each pulse is represented by the envelope 41 of FIG. 5. The pulses within a burst are spaced uniformly from each other. The spacing between pulses may vary from 5 nanoseconds to 10 microseconds.

Referring now to FIG. 3, the concentration of bursts 32 opposite the peak 28 of FIG. 2 can be expressed as a frequency of repetition. I presently prefer to adjust the FM capable generator 12 to have a maximum frequency of repetition in the range of 25 Khz to 100 Khz. I deliberately keep this range low to reduce the amount of heating caused by the microwaves. The wider spacing of the pulses 32 opposite the cutoff points 33 and 34 of FIG. 2 can also be expressed as a frequency of repetition and I presently prefer a minimum repetition rate of 1 KHz. I find that this low repetition rate, although in the audio range, does not disrupt the transmission of audio intelligence to the person 23. The aim, again, is to reduce the amount of heat transmitted to the subject 23.

Operation

Referring to FIG. 1, the intelligence to be perceived by the person 23 is introduced at the audio source 10 which may be a microphone for voice, or a tape player for music, instruction, etc. This audio signal is transmitted to the FM capable generator 12 and to the comparator 14. The comparator 14 compares the positive portions of the audio wave with voltage from the voltage standard 16 and when the audio wave instantaneously exceeds the standard voltage, the FM generator is actuated by the wire 17 connecting the comparator 14 and the FM generator 12. The FM generator 12 then sends a plurality of signals to the microwave generator 19 at each peak of the audio wave above the voltage standard.

This is shown graphically in FIGS. 2-5. The audio signal 27 of FIG. 2 exceeds the standard voltage 31 at point 33 whereupon the FM generator 12 starts emitting burst signals 32 at its lowest frequency of about 1 Khz. As time progresses past point 33 the voltage above the standard increases and the FM generator 12 responds by making the burst signals closer together until at peak 28 the maximum density of burst signals 32 is achieved, for example at a frequency of 50 Khz. The time duration of each pulse 40 (FIG. 4) is also controlled by a fixed adjustment of the FM generator 12 and for example the duration may be 100 nanoseconds.

The frequency modulated burst signals are delivered by FM generator 12 to the microwave generator as interrupted dc and the microwave generator is turned on in response to each pulse 40 and its output is delivered by coaxial cable 21 to the parabolic antenna 22 to project microwaves onto the head of a person 23. These microwaves penetrate the brain enough so that the electrical activity inside of the brain produces the sensation of sound. When the parameters are adjusted for the particular individual, he perceives intelligible audio, entirely independently of his external hearing organs.

Presently Preferred Quantities

As mentioned previously, I prefer that the standard voltage 31 of FIG. 2 be about 50% of peak audio voltage. This not only helps to reduce heating in the person 2 but also reduces spurious audio. This 50% is not vital and the useful range is 25% to 85% of peak audio.

The minimum burst repetition frequency (for example at time points 33 and 34) is preferably 1 KHz and the maximum repetition frequency is in the range of 25 KHz to 100 KHz, with the lower frequencies resulting in less heating.

The time duration of each individual pulse of microwave radiation is in the range of 10 nanoseconds to 1 microsecond as indicated in FIG. 5, with the shorter time periods resulting in less heating.

Control Of Power Output

As stated above, I maintain the power output of the parabolic antenna 22 within the present safe standard of 3.3 mw/cm2 (3.3 milliwatts per square centimeter). I control the power output by controlling the strength of the audio modulation. This results in a duty cycle of 0.005, the decimal measure of the time in any second that the transmitter is on full power. The peak power level can be between 500 mw and 5 w and at 0.005 duty cycle these peaks will result in an average power of 2.5 mw and 25 mw respectively. However, these values are further reduced by adjusting the audio modulation so that zero input produces a zero output. Since a voice signal, for example, is at maximum amplitude only a small fraction of the rime, the average power will be below the 3.3 mw/cm2 standard, even with 5 watts peak power.

Theory Of Operation

I have not been able to experiment to determine how my microwave system works, but from my interpretation of prior work done in this field I believe that the process is as follows. Any group of bursts related to the audio ek 28 of FIG. 2 causes an increasing ultrasonic build up within the head of a human being starting with a low level for the first bursts pulses and building up to a high level with the last bursts pulses of a group. This buildup, I believe, causes the direct discharge of random brain neurons. These discharges at audio frequency create a perception of sound. This process, I believe, bypasses the normal hearing organs and can create sound in a person who is nerve-dead deaf. However, this theory of operation is only my guess and may prove to be in error in the future.

Apparatus

The apparatus of FIG. 1 for carrying out my invention may include as a microwave generator Model PH40K of Applied Microwave Laboratories and described as Signal Source. The cable 21 connecting the microwave generator 19 and the antenna is RG8 coaxial cable by Belden Industries. The antenna 22 may be a standard parabolic antenna. The FM generator 12 has to be specially built to include the spacing function which is obtained by a frequency generator built into a standard FM generator.

I have described my invention with respect to a presently preferred embodiment as required by the patent statutes. It will be apparent to those skilled in the technology that many variations, modification and additions can be made. All such variations, modifications and additions that come within the true spirit and scope of the invention are included in the claims.

Reference Link

patents.google.com/patent/US4877027A/en

***

Mind Control and Behavioral Modification Technology

Reference: OHCHR "Directed Energy: Targeted Patents"

https://www.ohchr.org/sites/default/files/Documents/Issues/Torture/Call/NGOs/VIACTECAnnex.pdf

DARPA on your Mind

PMID: 15986543

Abstract

Applied science may once again play a decisive role in changing the face of armed conflict, and the rest of human affairs, by shifting the battlefield to our very brains. The national-security establishment--and particularly the Pentagon's **Defense Advanced Research Projects Agency**

(DARPA)--supports research at the intersection of neuroscience and national security that could ultimately enable authorities to do things like enhance (or muddle, or erase) memory, monitor crowds for individuals whose brain patterns correlate with aggressive behaviors, or control weapons from afar merely with thoughts. What are the dangers of such information falling into "the wrong hands," and are there any "right hands" for this kind of knowledge? Is any extension of human abilities justified by the need for government to protect its society?

Current research at the intersection of neuroscience and national security might one day produce weapons that literally boggle (or, if desired, enhance) the mind. This would give us unprecedented war-fighting superiority as well as a set of ethical dilemmas that could make genetically-modified-organism issues pale in comparison.

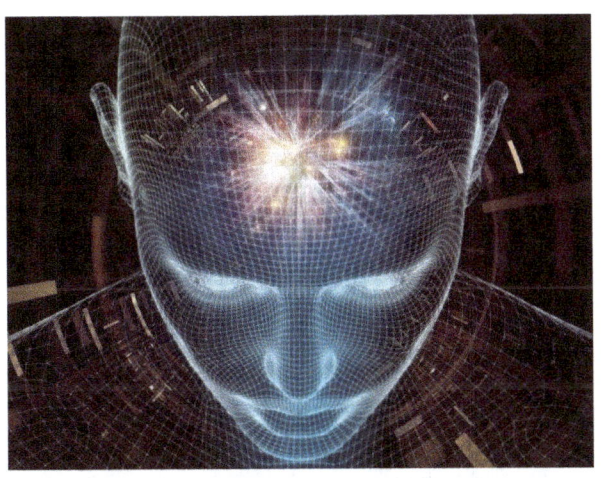

The Government Is Serious About Mind Control Weapons

DARPA, the Department of Defense's research arm, is paying scientists to invent ways to instantly read soldiers' minds using tools like genetic engineering of the human brain, nanotechnology and infrared beams. The end goals? Thought-controlled weapons, like swarms of drones that someone sends to the skies with a single thought or the ability to beam images from one brain to another.

***

USC Brain Computer Interface Research

A five-year, $11.25 million Multidisciplinary University Research Initiative grant has been awarded to a USC researcher who will lead a team developing brain-machine interfaces to enhance human decision-making.

The award given by the U.S. Department of Defense and the U.K. Ministry of Defense aims to connect scholars in neuroscience, machine learning and signal processing. The work will focus on developing new methods for modeling neural, behavioral and physiological data from humans in an attempt to understand the brain's multisensory processing.

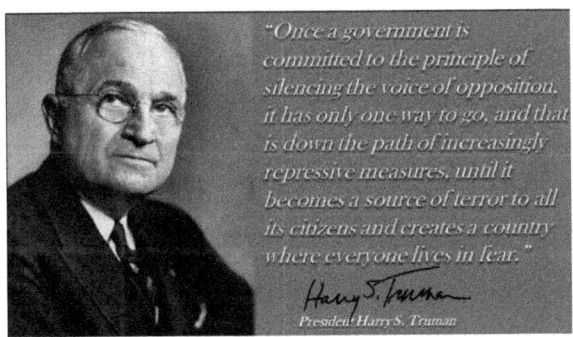

***

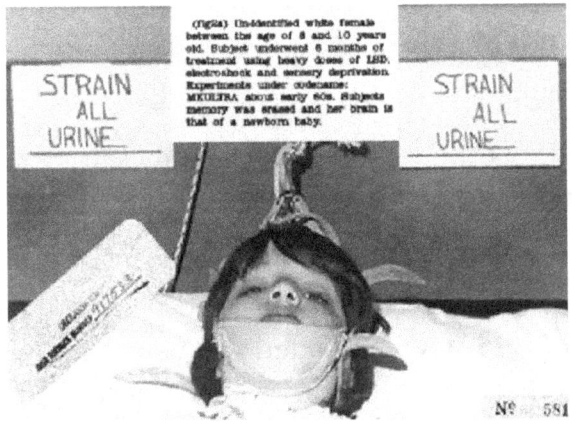

Mk Ultra History... *Lest We Forget*

That report prompted investigations by the United States Congress, in the form of the Church Committee, and by a commission known as the Rockefeller Commission that looked into the illegal domestic activities of the CIA, the FBI and intelligence-related agencies of the military.

The congressional committee investigating the CIA research, chaired by Senator Frank Church, concluded that prior consent was obviously not obtained from any of the subjects. The committee noted that the *experiments sponsored by these researchers ... call into question the decision by the agencies not to fix guidelines for experiments*

Following the recommendations of the Church Committee, President Gerald Ford in 1976 issued the first Executive Order on Intelligence Activities which, among other things, prohibited experimentation with drugs on human subjects, except with the informed consent, in writing and witnessed by a disinterested party, of each such human subject and in accordance with the guidelines issued by the National Commission.

Subsequent orders by the Presidents Carter and Reagan Administrations expanded the directive to apply to any human experimentation.

The U.S. General Accounting Office issued a report on September 28, 1994, which stated that between 1940 and 1974, DOD and other national security agencies studied thousands of human subjects in tests and experiments involving hazardous substances.

The quote from the study:

Between 1953 and 1964, this program consisted of 149 projects involving drug testing and other studies on unwitting human subjects with key plays from the Psychiatry,

Psychiatry and, for example of university connected studies, **Louis Jolyon West UCLA.**

Experimenters

- **Harold Alexander Abramson**
- **Donald Ewen Cameron**
- **Sidney Gottlieb**
- **Harris Isbell**
- **Louis Jolyon West**
- **Martin Theodore Orne**

Example of 'Alleged' subjects

Ted Kaczynski, an American domestic terrorist known as the Unabomber, was said to be a subject of a voluntary psychological study alleged by some sources to have been a part of MK Ultra. As a sophomore at Harvard, Kaczynski participated in a study described by author Alston Chase as a purposely brutalizing psychological experiment, led by Harvard psychologist Henry Murray. In total, Kaczynski spent 200 hours as part of the study.

Lawrence Teeter was the attorney for Sirhan Sirhan who assassinated Robert F. Kennedy, and he believed that Sirhan was operating under MK-ULTRA mind control techniques.

Theodore Kaczynski, also known as the Unabomber, was a participant in one of Henry Murray's experiments at Harvard where Murray's team bullied, harassed, and psychologically broke down participants. In reality, this tactic remains widespread across the nation reported as part of official mass targeting, patented mind invasive technology use. The tactic today is used today within the military COINTELPRO historically known as Nazi Zersetzung Organized Community Stalking psychological operations PsyOps, isolation, sleep deprivations, subliminal influence, combined with a recruited, nationwide network stalking.

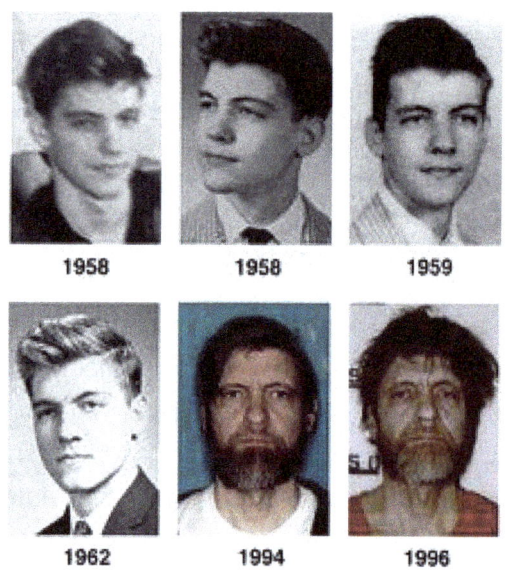

Was Ted Kaczynski Mind Controlled Puppet?

Hypnosis is an attention-focusing, consciousness-related procedure that consists of an induction stage and a suggestion stage (Kassin, 2004). In the induction stage, a person's attention becomes hyper-focused. In the suggestion stage, a person is open to suggestions made by the hypnotist. Hypnosis is sometimes used to treat phobias, stress, and pain

(Zimbardo, Johnson, & Weber, 2006). Evidence shows that those who are hypnotized will not comply with suggestions against their will (Wade & Tavris, 2000).

Individuals differ in their susceptibility to hypnosis (Kirsch & Braffman, 2001). Solomon Asch captured a historical context of hypnosis with a discussion of how interest in hypnosis had been the catalyst for social psychology's empirical research on more general suggestibility (Asch, 1952). The CIA's PROJECT ARTICHOKE used sodium pentothal and hypnosis on participants in search of more effective interrogation techniques (Select Committee to Study Governmental Operations with Respect to Intelligence Activities, United States Senate, 1976).

Tribute to Jose Delgado, Legendary and Slightly Scary Pioneer of Mind Control.

Neuroscientist based at Yale in 1960s controlled bulls, monkeys and humans with brain implants and envisioned a psycho-civilized society

Yale University And (Mind Control) Brain Chip Historic Study

***

Areas of the brain affected by hypnosis Brain-imaging studies show higher activity in the prefrontal cortex, parietal networks, and anterior cingulate cortex during hypnosis for suggestible subjects. These areas of the brain account for complex functions like processing emotions, learning, and perception and memory.

***

Stanford University Hypnotic Research

Study identifies brain areas altered during hypnotic trances

By scanning the brains of subjects while they were hypnotized, researchers at the School of Medicine were able to see the neural changes associated with hypnosis.

Columbia University Cia Gave Mind Tests At Columbia

The **experiments** were "designed to identify materials and methods useful in altering human behavior patterns," a **university** spokesman said." Study shows that people can boost attention by manipulating their own alpha brain waves.

Name	Frequency Range	Subjective Experience
Delta	0.5-3	Sleep
Theta	3-7	Imagery, suggestibility
Alpha	7-13	Relaxed awareness
Low Beta	13-18	Alert awareness
SMR	12-15	Sensory-motor rhythm
High Beta	18-30	Super alert, tense. High correlation anxiety when dominant.
Gamma	30 and up	Hyper alert, possible creativity

***

MIT Dream Manipulation Research

In a new paper, researchers from the Media Lab's Fluid Interfaces group introduce a novel method called *Targeted Dream Incubation* (TDI). This protocol, implemented through an app in conjunction with a wearable sleep-tracking sensor device, not only helps record dream reports, but also guides dreams toward particular themes by repeating targeted information at sleep onset, thereby enabling incorporation of this information into dream content. The TDI method and

accompanying technology serve as tools for controlled experimentation in dream study, widening avenues for research into how dreams impact emotion, creativity, memory, and beyond.

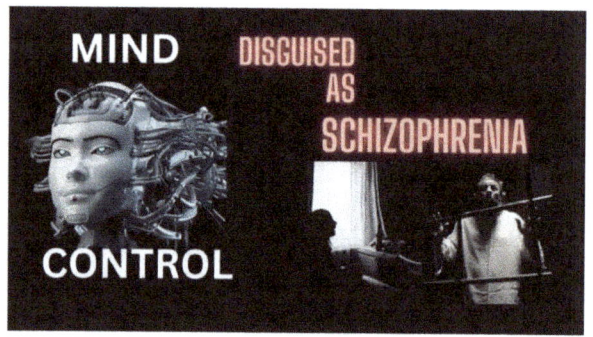

Deceived: Schizophrenic High-Tech Fraud

Renee Pittman's "Mind Control Technology" Blog

Reference Link

HOW IT WORKS: "Deceived – Mind Control – Schizophrenic High-Tech Fraud" a Creative Treatment of Actuality

The Targeted Individuals are now raising awareness to this crime by educating the public with websites, books, videos, and blogs.

Basically this is our main weapon in order to stop this crime, save our life and get compensated for the crimes committed against us that stole our lives from us.

Targeted Individuals public group

Frankly, I could not understand why the DEA was even around me. Alcohol was my hang up and it was purpose for the Arizona testing resulting from a DUI and this fact contrary to what this program who followed me there reported. There were no drugs in my system but there was alcohol. there was screening. However, realistically, many know it can make you vulnerable. Illegal substances were contradictory to my head and heart and both repelled it so there was no enjoyment that apparently others find. Again, this was a brief and I do mean brief, which is still used by corruption looking for ways to influence everyone and discredit me.

When my first book was published, immediately, federal agents created the image below and inserted it in Google Images around my books. My name here was no accident.

In gratitude I can tell you that, as my dad use to tell me, *Your normal mind is your best mind* even when in so much emotional pain that I did not want to think. I am happy intoxication bombarded my mind with a profound sense of guilt that was more powerful than any foolish, and I, thank God I was allowed to see the weaker side of me it too prepared me for going into battle. Sadly, as I look back,

in awareness, I realize that it was not just major depression that steered me in the wrong direction, as my awareness of the tactics of human experimentation evolved, combined with intense research and experiences that collided. It was, believe it or not, high-tech influence As I connected the dots, it become apparent that I and many others in my community was used as human experimentation since a child. Frankly, the last thing I would ever do is let is my three beautiful daughter's grow up thinking that their mother had this type of dysfunctional, revealing instead a human being able to conquer and self-correct during life's toughest storms.

I have often thought that the DEA had somehow, somewhere gotten wrong information about me and it likely from this program's overseers whom I document in Book VI as having been toying with me since childhood. Their involvement came to my awareness only after shopping for lipstick and a man, the same ones across the street today, followed me into and around the store as I shopped.

I remember when I bent down to look at items on a lower shelf, the man rushed up close to me, and bent down to see if I was picking up something, when I picked up a tube of lipstick to check out. Again, alcohol was the culprit for loosening my defense, resulting in putting major depression on top of major depression after my divorce. Dealing with a 15-year marriage to a policeman husband, then arriving in Los Angeles homeless after my vicious policeman husband used his job to have me arrested to get the kids and the house. I almost threw in the towel, crushed, and deciding to put some distance between him and I. Frankly I did not grasp as I fled the powerful emotional guilt of leaving without my children and this almost destroyed me, along with what I came to realize was ongoing human experimentation, every step of the way. On the weekends after the kids were sleep we would put on music, drink and relax. Today I hope he stopped, although it is likely he did not. I can tell you that a woman's body processes alcohol differently, so I could not keep up with him.

As an only child, with only my dad, in Los Angeles, I didn't have a support system while in Denver after my ex who came at me with everything he had, viciously, after I announced I wanted a divorce. During this time, I watched, broken and being verbally abused, him using the very job he had gotten through a relative of mine, who was recruiting for the police department on the military base we were living after two back to back overseas tours, 2 years in Korea and 3 years in Germany and 3 daughters. He said, I was not going to take his children away from him and thus the destructive tactics. When asked by the VA what happened, as shown, I documented this, and also how I became a target shown in one of many excerpts derived from my medical records. I was not only messed up emotionally but also gullible as a result of this. With me, I am the type to own my mistakes, and I will say this, and shout it from the highest mountain,

"THANK GOD I LANDED ON MY FEET!"

Thousands, upon thousands become lost and self-entrapped, in situations such as this that crosses all barriers, all races, poor, middle class and affluent and never make it to the light.

> MD Psychiatry
> 03/05/2012 ADDENDUM   STATUS: COMPLETED
> Patient interviewed and discussed with team. Patient expressed elaborate and intensely held beliefs that she is being the target of the FBI in an attempt to use microwave technology to prevent her from publishing a book about her paranoid delusions. She states that her ex-husband is a police officer and that the police in her city are investigating her at the request of the FBI. In the investigation the police asked her ex if she were sexually traumatized in the military and the ex stated she was not. She also states that gang-stalking is being employed by the police to make her stop writing her expository books. She states her downstairs neighbor is participating in gang stalking and has a microwave generating device on top of a wall unit she has seen through his window. He uses the unit to send microwaves into her apartment. The microwaves have caused her hip and ankle to not heal properly and she states it causes her nails to wrinkle. She has learned that these are side effects of being

I had not lived in Los Angeles for over 15 years joining the military young, then working transferring to positions as a DOD employed civilian working on military bases. I did not understand, and was not prepared for professional people, I had grown up with, adapting to using substances recreationally.

What I realize today as they try to silence me, using the stereotypical drug trope for Black people plays in discrediting and mobilizing Community Volunteers for official, Organized Community Stalking. You can't blame those told this lie about me. No one wants drugs in their community myself included. I took control of my life in 2007, and got a grip.

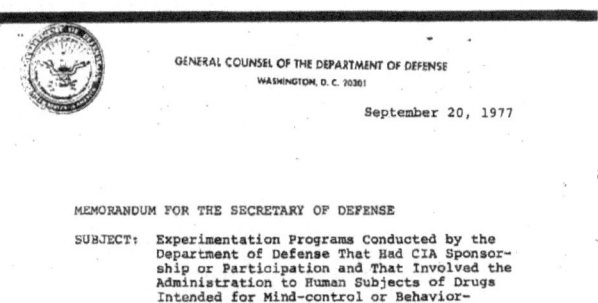

Yet federal agents are still influencing personnel and likely this woman, the author of *Violence and Gender* where she focuses on me by putting me in the league of mass shooters and half-truth regarding Myron May, who she fails to mention was a heavily target and destroyed African American District Attorney.

One thing I am in total agreement with, this author and anyone reporting that Targeted Individuals are harming self and others.

Many across our nation, who outnumber those who take this kind of drastic and horrific action, full recognize that this is mental illness, which can be and is being created by advanced mind invasive technology, magnifying *Crazy is as Crazy Does!*

Thousands fighting the good fight also understand how everyone can be and are lumped the same category. This is completely understandable as well. However, there will always be weakness in humans and people who snap, give up or give in and throw in the towel for other reasons or who have become monstrously angry. This is why the Targeted Individual's Blue-Ribbon logo specifically serves as a reminder for those suffering to not bend or break for this reason.

The fact is, if these people are targeted for human experimentation as many say they are, they had been living under a tremendous, 24/7 high-tech mental, with emotional degradation and psychophysical assault by this program and the careless officials at the helm of these space-based systems and devices with their lives deemed useless by socio and officiated psychopaths, and in the name of advancing behavioral modification technologies which is the essence of mind control.

I can guarantee, none of the highly credible individuals, exposing this program, are far from agreement with this type of heinous deadly character. There is no excuse and if excuses are made, the person is someone to be weary of. The truth is the most powerful of all weapons and this program knows it!

Basically, this is also what I told Assistant District Attorney, Myron May, a person who reached out to not only me and several others on social network at the same time, known for exposing this program and he did so a mere 6 days before he took the action he did. He ultimately sent out 10

Certified Mail packages to 10 individuals, as he put it hoping that his story never dies with him. This was a young, by all account brilliant young attorney on the rise! Until he became monstrously targeted.

His story, *The Targeting of Myron, Asst. DA Pushed Over the Edge* is in both printed and eBook formats. Frankly speaking, I was angry that he used me this way, knowing in advance what he decided to do, that the book almost did not get published. It was not until those involved, federal agents, and the military mind control and cops in Florida tried to connect me to the decision he made insinuating I was a motivator to a complete stranger.

The Targeting of Myron May, Asst. DA Pushed Over the Edge by Renee Pittman

A LinkedIn search of Ms. Sarteschi reveals her connection as an Educator of all levels of law enforcement, and connected at the highest level of leadership. I see nothing wrong here and recognize so-called people such as herself would never be privy to secret intelligence activities and her goals while throwing others under the bus, appear to be naively sincere. The fact is people like her are this program's best friend and advocate. For example, the Association of Psychiatry unified with Big Pharma has been involve in mind control studies since the MK ULTRA's 20-year program ran in the U.S. and led by then former President Donald. Ewen Cameron, detailed on Wikipedia who specialty was Psychiatry and Mind Control.

Donald Ewen Cameron		
	Born	24 December 1901[1] Bridge of Allan, Stirlingshire, Scotland
	Died	8 September 1967 (aged 65)[1] Lake Placid, New York, U.S.
	Nationality	Scottish-American
	Scientific career	
	Fields	Psychiatry, Mind Control

Cameron, circa 1967

There are 80 across our nation and growing, of which many highly credible individuals, including Karen Melton-Stewart, a targeted 28-year veteran of the National Security Agency, reports are factually overseeing this program, with the Department of Homeland Security the military and intel agencies. It should be noted that routine, written NSA psych evaluations, evaluated independently, over 20 yrs. While employed by the NSA, confirmed a clean bill of mental health. This was until she started reporting corruption, and subsequently, the high-tech human experimentation program resulting in her placed on the list and Targeting.

Many of Ms. Sarteschi's LinkedIn posts publicize job opportunities for Fusion Center recruitment which is not wrong.

However, if you asked who is officially spearheading this program on an administrative level, fusing civilian and military agencies together, hands down, as shown by the flyers create by Karen Melton-Stewart, it is nationwide connected Fusion Centers. In my case, the overseer is the Los Angeles County Joint Resource Intelligence Center (Fusion Center) in Norwalk, California.

Frankly the only issue I have, on a personal level, with Ms. Sarteschi is lumping me in with people who have snapped when I stand with many, thousands, highly credible, who have stood the test of time and stood our ground staying the course altruistically dedicated to the truth. For me, now entering 18 years of relentless focus and efforts to set up destroy, entrap and shut down. Violence destroys everything we have work for hoping awareness will save lives!

Karen Melton Stewart @ka... Apr 16

#NSAbadge - Held Top Secret clearance for 27 years until I asked the NSA Inspector General to investigate something.
pic.twitter.com/LTle14wxgB

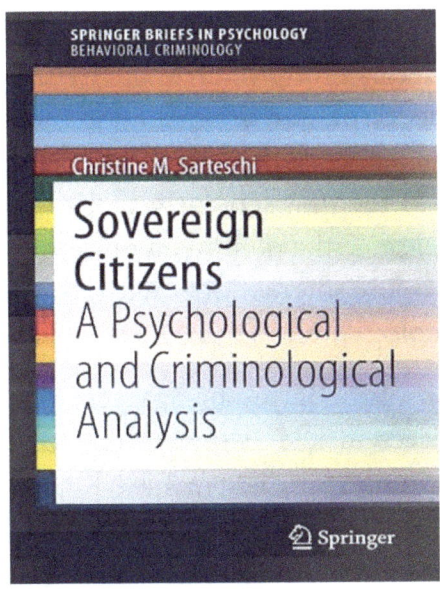

Sovereign Citizens: A Psychological and Criminological Analysis (SpringerBriefs in Behavioral Criminology) 1st ed. 2020 Edition

by Christine M. Spartech (Author)

3.2   3.2   out   of   5   stars   3   ratings

This brief serves to educate readers about the sovereign citizen movement, presenting relevant case studies and offering suggestions for measures to address problems caused by this movement. Sovereign citizens are considered by the Federal Bureau of Investigation (FBI) to be a prominent domestic terrorist threat in the United States, and are broadly defined as a loosely-afflicted anti-government group who believes that the United States government and its laws are invalid and fraudulent. Because they consider themselves to be immune to the consequences of American law, members identifying with this group often engage in criminal activities such as tax fraud, paper terrorism, and in more extreme cases, attempted murder or other acts of violence. Sovereign Citizens is one of the first scholarly works to explicitly focus on the sovereign citizen movement by explaining the movement's origin, interactions with the criminal justice system, and ideology.

Another example of Sarteschi's perceptions are revealed in the Science Direct article below which could likely place Social Workers in danger. The fact is a lot of highly educated people feel the same way about Government. It is not just the so-called underclass and again, not all perceived this way violent. What we witnessed at the Capital was a tiny percentage of 330 million people in this country.

> Sovereign citizens: A narrative review with implications of violence towards law enforcement
>
> Christine M. Sarteschi

While her efforts appear to be both honorable and biased consideration must also be given to the fact, as reported in The Guardian dated January 8, 2024, and those she serves. She is likely aware of the fact that police in the US killed at least 1,232 people last year, making 2023 the deadliest year for homicides committed by law enforcement in more than a decade, according to newly released data.

Frankly if Ms. Sarteschi were aware of Federal, state and local law enforcement, Fusion Center, Military Intelligence involvement in mind invasive technology approved for research, police investigations, and bioweapons for crowd control, she would be no longer effect in her criminology profession.

Ms. Sarteschi has several controversial opinions shown below:

EXPERT OPINION

Social Workers Cannot Yet Replace the Police

"Until police officers are better trained in mental health crisis intervention techniques, many mental health calls are likely to need a both a social worker and a law enforcement officer, say Christine M. Sarteschi, an Associate Professor of Social Work and, .Criminology at Chatham University in Pittsburgh, PA and Daniel Pollack, MSSA (MSW), Attorney and Professor at Yeshiva University's School of Social Work in New York City"

Reference Link

https://www.law.com/texaslawyer/2021/03/22/social-workers-cannot-yet-replace-the-police/?slreturn=20240116234109

***

Social Workers Won't Replace Police

Anytime Soon - WSJ

One study of nine cities found that mental-health calls make up less than 2% of 911 calls. It's nice to have someone do wellness checks. But sending social workers into low-stakes situations is unlikely to reduce police shootings of the mentally ill. Jul 20, 2022

Lastly, she appears to mock my books as self-published, with multiple reviews versus her *Sovereign Citizen...*" book with three reviews as of 2020. Over the years, this has also been a discrediting tactic of the federal agents around me influencing people that nonfiction books, derived from open literature evidence are not credible.

The fact is self-publishing is one on the best moves an author can make and this especially true with nonfiction and I have complete control and can make updates.

Because, Ease of Access to Publishing Platforms: Self-publishing has gained significant popularity over the years due to the ease of access to publishing platforms and the ability for authors to retain creative control and higher royalties.

So here is the list of 10 most famous self-published authors who dared their dreams and made it their realities:

1. E.L. James
2. Virginia Woolf
3. Shubham Shukla
4. Ashish Bagrecha
5. D.H. Lawrence
6. Anubhav Aga rwal

7. **Rupi Kaur**

8. **Amish Tripathi**

9. **Ashwin Sanghi**

Generally, when you think about self-publishing you think about people that are new to this field and who can't afford working with an editor or a large publishing house. But the truth is you can find celebrities and famous people that chose to self-publish and make it big. This is good to note, because every author can publish a book themselves without relying on external sources.

Stephen King is a well-known author and ended up self-publishing a book on his own in 1960. *Named People, Places and Things*, this book was published by Triad and Gaslight Books, who is his own publishing company. He chose this route because he wanted to get the book in front of as many people as possible.

The fact is, of which I document in my first book, *You Are Not My Big Brother…* before I decided that Remote Brain Targeting with open literature should lead the series for credibility, I had a publisher until federal agents bullied the publishing company out of publishing. It was either give up or do it myself. Today, I am glad I did because I retain complete control over everything and can also do updates to manuscripts if and when needed

TODAY FEDERAL GOVERNMENT NON-CONSENSUAL DEADLY EXPERIMENTS *CONTINUE* IN YOUR NEIGHBORHOOD

Directed Energy Weapons are being tested in your neighborhood <u>non-consensually</u> under the auspices of **DHS / FBI / Fusion Centers** under the guise of *long-term* "Terrorist Watch List investigations". Complicit neighbors are poisoning you and themselves for under-the-table tax dollars. Electromagnetic energy causes Cancer in *all* biological life. The unborn and children are most at risk. DNA damage is likely.

<u>HISTORY of COVERT ABUSES</u> In the 20th Century:

- Project MK Ultra, Subproject 68
- Mustard Gas Exposure on soldiers in gas chambers
- Deadly Chemical Sprays over U.S. Cities
- Intentional Exposure of Guatemalans to Syphilis, left untreated
- Manhattan Project: Plutonium & Uranium Injections
- Agent Orange Injections Given Prisoners
- Operation Paperclip - WWII Nazi Scientists given US Citizenship
- Puerto Ricans Injected with Cancer
- Department of Defense Exposes Blacks to Extreme Radiation - 7500 x normal X-ray
- Operation Midnight Climax - CIA Brothels in NY & SF exposed Clients to LSD
- U.S. PACIFIC ISLAND Radiation Fall Out / PROJECT 4.1
- TUSKEGEE, AL 400 Black Sharecroppers Exposed to Syphilis, left untreated.

SEE: EverydayConcerned.com ; FreedomForTargetedIndividuals.org
http://wariscrime.com/new/the-13-most-evil-us-government-human-experiments/
Karen Melton Stewart, retired NSA Intelligence Analyst

ATTENTION LAW ENFORCEMENT

The "Suspicious Person" you are ordered to
- spread rumors about,
- keep an eye on,
- overtly harass,
- deny equal protection under the law,
- treat *as If* they were "crazy",
- *falsify* charges against "for the good of the community"

Is NOT a

- pedophile,
- traitor,
- terrorist,
- criminal,
- crazy person,
- prostitute, etc.

They are a RANDOMLY CHOSEN SCAPE GOAT to FALSELY present as a threat, in order for useful idiots to harass and vilify so that the Local Fusion Center and YOUR SUPERIORS can justify the expansion and bloated budget of an unneeded POLICE STATE at the expense of INNOCENT PEOPLE and YOUR COUNTRY / LAND OF THE (not so) FREE.

YOU ARE BEING PLAYED FOR FOOLS by corrupt, greedy enemies of America, hell bent on subverting freedom and justice and doing so by destroying thousands of innocent lives, so THEY can run America like a dictatorship.

Is THAT why YOU joined the police force? WISE UP. REPLACE CORRUPT SUPERIORS.

FUSION CENTER INFRAGARD (useful idiots)

FUSION CENTERS
INFRAGARD
(Useful idiots aka the Red Terror)

Definition:

GULLIBLE, GREEDY *TRAITORS* HELPING THE DEEP STATE *SECRETLY MURDER*

- PATRIOTS,

- PEOPLE OF INTEGRITY, &

- INDEPENDENT *THINKERS*

... *UNLIKE THEM.*

🔥😡🔥😡🔥😡🔥😡🔥😡🔥😡🔥😡🔥

Again, this program is suppressing motivated human experimentation and covert murder, watching and monitored in real time. In fact, with the setup around me, as of February 11, 2024, the cops and military personnel watched me completing this manuscript, and again and again began another attack to my abdomen.

This is again, an official hope that they can create synthetic colon cancer. Contrary Sarteschi ridiculing of anyone reporting this program they can be set up in our communities for up close training, and worse, and use of military weapons, and weaponized drones.

If I end of with any type of cancer, whether it be colon, breast, uterus, a brain tumor, etc., heart attack or stroke, it was a directed result of ongoing official efforts to silence the truth. This includes damage to my knees.

The mental illness tag has not stuck to me for them and they are monstrously angry, while sitting behind these computer systems watching and reading what I am many others are reporting about them with me specifically creating this book.

Christine Sarteschi is also wrong about me being diagnosed with schizophrenia.

As shown, the VA hospital note documented that I do not have the characteristics of schizophrenia although military COINTELPRO have tried to make this label stick to everyone exposing the truth of patented beamed communications and sadly effective with many for discrediting.

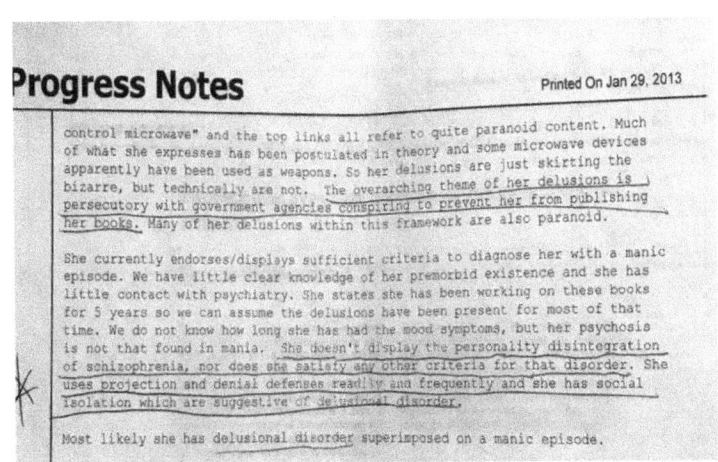

It should be noted that I have never taken antipsychotic medication for any type of psychosis. For PTSD the VA prescribes Prazosin and Hydroxyzine. Prazosin was once used to stabilize blood pressure with the resulting benefit of a peaceful night's sleep. Hydroxyzine is the VA's recommendation for anxiety and is similar to Benadryl.

Frankly, these medications prescribed for PTSD are what saved me and also became a beneficial blessing for me enabling me to write books with a good night's sleep and coping with the anxiety this program tries to create to push targets into a negative reaction for both discrediting and strategic diabolical silencing.

```
BUN:    6 mg/dL      SEP 14,2012 03:58
GLU:    121 mg/dL  H  SEP 14,2012 03:58
Hemoglobin A1C:  6.0 %  H  MAR 29,2012 11:38


HOSPITAL AND ACUTE REHAB UNIT COURSE:
Patient was transferred to the Acute Inpatient Rehabilitation on 9/17/12.
Patient had a fall on 9/21/12 outside of building near building 212 2/2 not
being able to stop the one of the brakes while going fast down a hill to catch
the shuttle. b/l XR hip: no acute fx/dislocation. Patient did not complain of
increasd pain.
Additionally, patient exhibit delusional thoughts about government conspiracy
for her writing her book (recently published - she has it at bedside) and
punishing her with microwave implanted in her knee.  Patient is highly
intelligent and is very articulate. Psychiatry consult 9/19/12: "Pt stable and
shows no mental status signs or symptoms of PTSD, mania, or cognitive
impairment. Nursing and housestaff indicate she is cooperative and progressing
well. Recommend no changes in medications or treatmnents at this time.
May d/c home when ready. No hold indicated."

The patient made significant improvements in ROM, Strength, FIM's and mobility.
Pt successfully completed an acute interdisciplinary inpatient rehabilitation
program and is appropriate to be discharged with appropriate follow-up
appointments.

I. MEDICAL/SURGICAL ISSUES
MEDICAL DIAGNOSES AND PLANS:
53 yo F w/ h/o HTN, HL, PTSD, and obesity s/p L THA on 9/13/12 due to severe OA.
Pt tolerated the procedure well without any complications. She was placed on
lovenox for DVT ppx.
```

Sections of high-level government agencies, connecting federal, state and local police who, Post 9/11 unified with our military under one umbrella, are on the hunt for so-called Domestic Terrorist, and have used this label placed on anyone, whistleblowers, activist, political dissidents, and any awakening to being used for human experimentation then becoming as a threat.

Once this label is attached, it opens official human experimentation for mind invasive, mind reading, and patented verbally beamed harassment technology where these agencies use on targets for 24/7 which is key in discrediting people they recognize as a threat.

Lastly, regarding Ms. Sarteschi stating that I checked myself into the hospital for mental health evaluations. This is 100% true. This happens to everyone. Initially you actually think you are going crazy when this program rears its ugly head and on top of this I was severely depressed which can also be synthetically assisted by manipulation of brainwaves and on a massive scale, believe it or not.

Actually, again, high-tech brainwave manipulation is key for behavior modification and a capability of high-tech operations, including, portable systems used from neighboring locations, such as apartments above you, and military operation centers nationwide, and police connected to electromagnetic energized space-based systems. There is no shame here. It is a firm belief that *Crazy people, do not think they are crazy!*

The first thing any logical person would do is seek help!

The fact is the horrific farce was going well for them until various tactic revealed there is something else monstrously at play around many. I woke up when I realized that these seedy, perverted personnel at the helm will use the sexual stimulation patent, noted in other books as well, by Hendricks Loos, to sexually stimulate women, unaware, who think they are alone in the privacy of their home. The patent is so powerful that you have no choice but to try to relieve yourself resulting in an unknown to you, real-time monitored show for the real crazies.

The fact is none of these individuals will never back track on what they are promoting and teaching. admit they are wrong about covert human experimentation, or their education is flawed and that this program does exists. As a result, they become, a powerful ally to the agencies involved backing up silencing of the truth with the mental illness tag.

The issue here is the violence that is a result of human experimentation, and relatively unknown to the general public. This demands exposure to save lives. Nor, will the military or federal agencies, and all involved at this level admit this program or involvement.

Truth reveals official personnel as far less than honorable and in the case of the underlings, the truth deprives them of desperately coveted importance wanting to be someone, no matter how with government sanction.

Image created by Targeted Justice, Member who was a NASA engineer and now targeted himself.

Below is the District Court filing filed in Arizona, Sarteschi mentions in her article. At that time, I did not know who the real culprits were used an online example of a District Court filing submitted by someone, a target experiencing the same experiences. My fine point is being candid with nothing to hide.

Later the second District Court filing was filed in California after realizing who was using me as a human guinea pig, who came into the open when I started publishing books, blogs and a website revealing my plight and suffering along with those of many nationwide.

Case 2:09-cv-01659-JAT Document 1 Filed 08/12/09 Page 1 of 11

RENEE PITTMAN MITCHELL
8450 N. 67th Avenue, Apartment # 1078
Glendale, Arizona 85302

IN THE UNITED STATES DISRICT COURT

FOR THE DISTRICT OF ARIZONA

RENEE PITTMAN MITCHELL

    Plaintiff,

Vs.

CV-09-1659-PHX-JAT

National Security Agency / Central Security Service
Department of Homeland Security
Department of Justice (Federal Bureau of Investigation and /
Drug Enforcement Agencies of California and Arizona)
Anaheim California Police Department
Los Angeles California Police Department,
Signal Hill, California Police Department
Scottsdale, Arizona Police Department
Oakland, California Police Department
Glendale, Arizona Police Department

And

UNKNOWN OFFICIALS
NSA/CSS, Dept of Homeland Security. FBI, DEA and local police departments

    Defendants

COMPLAINT FOR DECLARATORY AND INJUNCTIVE RELIEF

Ms. Mitchell hereby files this complaint for declaratory and injunctive relief as a result of violations of her First and Fourth Amendment Rights. As grounds therefore, Ms. Mitchell respectfully alleges as follows:

PRELIMINARY STATEMENT

1. This lawsuit challenges the constitutionality of a secret government program (hereinafter "the Program") to use spy satellite and provide X-Ray Vision technology to

neighbors for viewing inside the residence an Ms. Mitchell during normal activities and more specifically inside the privacy of her bathroom using both technologies. The National Security Agency / Central Security Service ("NSA") launched the Program in 2001 and the President of the United States ratified it in 2002.. The Department of Homeland Security launched a Pilot Program for local police departments in 2007 that concluded on June 23, 2009.

2. The plaintiff is an American citizen, who is a 100% Service Connected female veteran of the United States Army diagnosed with Post Traumatic Stress Disorder (PTSD) as a direct result of Military Sexual Trauma whose further mental health and well-being has been further jeopardized, and traumatized as a direct result of the spy satellite invasion inside her home and specifically her bathroom and X-Ray Vision technology provided to her neighbors for viewing inside her residence and spy satellite used in any and all public locations she frequents to include derogatory comments made in attempts at humiliation and degradation while in these locations. The Program also violates her right to privacy during phone conversations regarding legal and mental health and medical doctor appointments.

JURISDICTION AND VENUE

3. This case arises under the United States Constitution and the laws of the United States and presents a Federal question within the Court's Jurisdiction under Article III of the United States Constitution and 28 U.S.C. 1331. The Court also has jurisdiction under the Administrative Procedures Act, 5 U.S.C. 702. The Court has authority to grant declaratory relief pursuant to the Declaratory Judgment Act, 28 U.S.C. 2201 es seq. The Court has authority to award costs and attorney fees under 28 U.S.C. 2412. Venue is proper in this district under 28 U.S. C. 1391 (e).

COMPLAINT

4. In 2006, I became aware of a wiretap on my cell and house phone and electronic and spy satellite surveillance in my vehicle and inside my apartment. I was untreated for symptoms of PTSD and self medicating. In March, of 2006, after completing a court ordered substance program, I decided to leave California for Scottsdale, Arizona for a fresh start and new beginning. The investigation followed me and it was during this time I became aware of their presence inside my bathroom as I undressed, dressed or bathed. (I have numerous letter of complaint dating back to 2006 to Judges, U.S. Attorneys, Inspector General, etc., explaining what I was experiencing.) I was horrified as I innocently tried to relieve myself sexually in the bathtub as I was not then and today am not sexually active. I heard laughter coming from nowhere and then hush voices as I listened in shock, horror, disbelief and embarrassment to laughter and a man telling another what I was doing while giggling. I felt a wave of nausea and was gripped with fear and didn't know if what I was experiencing was real or a psychotic episode. I was 5 months sober when I move to Scottsdale and relapsed 8 months later. I also noticed at that time that whenever I moved from room to room inside my apartment, the neighbor

upstairs did the same in sequence with me and it sounded like he was dragging something which I later found out was an X-Ray Vision Machine. Comments were literally coming from the walls such as "She's got to go or She's going to jail," and "You won't be attending school" as I discussed with the Veterans Administration about taking classes to become self sufficient and I continued to hear lewd and lascivious sexual comments. This degradation contributed to my relapse and I started having nightmares and stayed away from my residence as much as possible. In December of 2006, I got a misdemeanor DUI. I stayed around the house mostly from then and was reluctant to go out. I was constantly badgered with threats coming from the walls and laughter and sexual comments. In March of 2007, after an unbearable week of just plain terrorism, I got drunk feeling so very hurt desperate for it to stop, severely depressed and humiliated that these people thought so little of me and treated me in such vile, degrading manner (as I exited the bathtub one night, one of them said "Yea!" as I stepped naked from the bathtub then I heard laughter.) I slipped on clothes ran out of my apartment, drove off then decided that it was foolish to leave and that it was best and safest to go back home. I had a near fatal car accident that night. As I tried to turn around I got lost and exited the freeway trying to find my way back onto the freeway. I was going about 30 miles a hour when I hit a brick median on a unlit road. I had the first surgery on March 18, 2007 during which time an external fixation device was place and the second in April to remove it during which time several screws, pins and titanium rods were place in my leg restructuring my ankle.

5. I was admitted to the Phoenix VA Hospital Mental Health Ward in April and May of 2007 as a direct result of what I was experiencing. I left the Phoenix VA Mental Health Ward for the Prescott VA Veteran Substance Abuse Program. I lasted there 2 weeks before I was irregularly discharged. My mental and physical state was more than they could handled or that I could handle myself. I was messed up. I returned to Los Angeles, California where I relapsed for a month before again being admitted to the Mental Health Ward of the West Los Angeles VA Hospital in July of 2007. Afterwards I went into intense therapy and treatment for PTSD with a dual diagnosis of substance abuse. After over a year in therapy, I found a efficiency apartment in Signal Hill a suburb in Long Beach, California. It was during my stay at the apartment in Signal Hill that I came under the most vile, disgusting, egregious, unethical, immoral, degrading behavior imaginable. One month after I moved there, in July of 2008, I noticed that the apartment vacant apartment next to me had been set up as a possible surveillance apartment with the X-Ray Vision technology provided to the apartment manager. I could hear someone positioning the technology against the wall as I watched TV. I went to the local police department to complain. I complained to the landlord who told me that he worked for the police. The worse however was as I sat reading and heard a creaking noise across the ceiling as I sat in a chair. There was no apartment over me. I noticed as I got up and moved from the chair, whatever it was crept across the ceiling and positioned itself directly over my head. If I then moved to the bed, it did the same. The most devastating thing was when I noticed that each and every time around my normal bathing time, I would hear it enter the apartment and position itself directly over my shower in the bathroom. This went on daily, 24/7 and continues without fail to this very day. They would and are saying things

for me to hear them such as "Did she take her shower yet?" Or "Damm, did you see her breast (38DD) and their real" or "she's cute." This goes on day after day, morning, noon and night until finally the women started to laugh at me and saying "Oh she thinks she's cute" or "Oh, she's not so cute and did you see her stomach." "Ugh!" She looks like she's pregnant." I have a 13 cm fibroid tumor. For many months I would bathe in the dark and still do as they continue this behavior today here in Arizona and every locations I have lived. They began to stalk me and centered around my head inside the wall as I tried to sleep at night and still do to this day and would say, "She's going to jail or We've got her" and "She's got to go!" over and over again, 24 hours a day, 7 days a week from inside my apartment and the background of my cell phone as I speak with anyone. They watch me change my sanitary napkins and comment "Ugh!" all the while professing that they are going to come out and arrest me "We got her now, Let's get her, She's going to jail." I began calling the DEA in September of 2008. During the first call from my wiretapped cell phone, I heard a man say "See I told you she is smart" in the background. They after a while then switched, after realizing that I was drug free, to "She lied about being raped." "Yea, it true, she lied about being raped." This is absolutely not true and I cannot begin to tell you how it feels when something so devastating and traumatic and life changing in your life is questioned and negated in such a manner and you are vilified when it was a documented event and fully investigated and to my knowledge the VA told these people that they were not pressing charges as I had told someone after it happened and there was evidence in my military personnel records to substantiate my claim. As a direct result of them watching me naked, one of the cop's came out while I lived in Signal Hill, pulled out his penis trying to seduce me while wearing a listening device and interrogating me about the rape. After I told him who I suspected he was, he became enraged and told me that I would continue to be "stalked" as he put it. (I heard him tell his partner one night before he came out, as he watch me make a sandwich in my kitchen, that he was going to "F____" me as he put it.) I continued to call the DEA and complain each morning. I could hear them in the bathroom so much that they actually stopped answering the phone. When I blocked my number and called them back one morning, a woman agent answered and angrily said "We turned that case over to the FBI" I said "What case, mines?" (I still was unsure) she said "Yes, why don't you call them"

6. In January 2009, my middle daughter became gravely ill in Oakland California. I moved to Oakland to tend to her while she was healing. The surveillance continued with threats, degradation, harassment and terrorism and the elderly neighbor upstairs and brother following me from room to room again with the X-Ray Vision technology following my every move As my neighbors still do here in Arizona on both side of me. The neighbor upstairs was doing this also but moved out. I called the cops on her several times.

7. In April of 2009, I decided to leave California for a spiritual school in Albuquerque, New Mexico. I took the battery out of my cell phone. I had been receiving threats of "You are not going anywhere" and "You will never make it to Albuquerque" and "You're going to jail." I left Oakland around 3:30 p.m. on March 19, 2009. Around 3:30 a.m. I ran out of gas and put the battery back inside my phone to call for a 911 operator, tow

truck and gas. As my car began to stall on the side of the road, in the darkness in the mountains just outside of Buckeye, Arizona while diesel trucks raced by, I talked to the 911 Operator, the cops in the background of my phone said "Got cha!" They then amplified and echoed my voice loudly as I tried to talked to the 911 Operator so that I could not hear her. I started screaming at them threatening a law suit. I could not hear the 911 operator and she was having trouble understanding me due to them. I was so upset that the 911 Operator asked me who was I yelling at. I told her no one not wanting to appear like a nutcase and not get any assistance. I heard the older female cop who I have become familiar with say in the background of my cell phone, "Don't do that you are just going to put more money into her pocket." I took this is referring to my legitimate VA compensation and Social Security Benefits which again the Veteran's Administration and the Social Security Administration did a thorough investigation prior to approving. Before I left Oakland, I was informed by the VA that it was not them investigating me and I later learned that these individual were angry that the VA told them that they were not committed to charging me with anything as I had told someone after it happened albeit my bitter ex-husband who to this day blames me for not getting promoted on the Denver Police Department. I received gas and got back on the road. About 3 miles later I was pulled over. This is when I found out that charges had been put on the book for the March 15, 2007 car accident 8 months after I had left the state. I also heard the arresting officer say "Dang, they really want this her," first at the arrest sight then while in processing at the Estrella Jail. When I questioned the officer about what exactly he meant after hearing him a second time say it, his partner answered saying that was just confirming the warrant. I was a spectacle at the jail as jail personnel whispered "The Feds are after her." I stayed there until April 15, 2009. I had dissociated I am told and didn't regain any sense of reality until around that time. I had saved enough money to bail out on the 1st of March but was psychologically void and numb. While there, the surveillance continued inside the jail with jail personnel and also via satellite as evident when I walked into the bathroom to change my sanitary pad and heard a man say from nowhere say, "Is that her" then "Yea" and "Damm, she is nasty." I could hear them around my bunk as I talked to other inmates saying "Damm she good" because I did not say anything beneficial for them. This was not by manipulation. I simply had nothing untrue to say. Whenever I used the phone I could hear them as usual in the background of the line. The jail personnel commented that the Feds were after me as if I was some sort of freakish fascination. When I filed a form to try and get an attorney as I knew no one here in Arizona, and was passing through to get my furniture out of storage from 2007, it came back at the top was annotated at the top "To be detained." I questioned this with no explanation. Finally as I tried to bail out a Sergeant had to come and tell his personnel that the Feds did not have the authority to detain me and to let me out. While incarcerated, at the Statutory Hearing on March 26, 2009, while I sat with the other inmates, I heard a young male Mesa Public Defender telling one of the Sheriff's guarding us that the Feds were after me while smiling smugly like he was involved in something really big as the sheriff tried not to look over at me. When I got to speak with the public defender Ms. Wallace, she told me what I was charged with and that they said that drugs were in my system and that I hit a brick wall and would have to pay $100,000 in restitution among other things. I immediately began to explain to her about what has

318

been happening to me these past 3 years. She asked if I have any psyche diagnosis, I truthfully told her yes that I am diagnosed 100% Service Connected for PTSD. She told me to have a seat that she wanted some other hearing for me and I overheard her tell the sheriff that something was not right with my case and she did not feel comfortable with that. I also observed a male trying to pass her a brown folder and watching me closely as I approached her for our conference. I did not know what she meant regarding another hearing. I did not have drugs in my system and I did not hit a brick wall and the only time I said this was after I left Arizona and returned to California and tried to explain what happened to me to family and friends describing metaphorically about hitting a brick wall psychologically and my urgent desire to get my life in order. As I bailed out, the surveillance picked up immediately inside my bathroom and apartment with the same lewd and lascivious sexual comment along with the usual threats, harassment, terrorism and degradation.

8. They have listened as I talk legal people about the DUI and their involvement in my life and they watch and listen as I meet with mental health personnel in therapy at the Arizona Carl T. Hayden VA Hospital. The young public defender in Mesa showed up 2 months later at the Mesa Courthouse as I was trying to deliver copies of the numerous letters of complaint I have written to senior government officials over the years to the Judge in the Mesa Courthouse who at the time I erroneously thought was over my case my however I later learned that my case had been transferred to downtown Phoenix. He aggressively insisted on delivering them for me then later I heard the agent commenting inside my wall say "Man, she's got letters dating back to 2006." When I called the court to inquire as to whether the judge had ever received the letters I gave to the Mesa Public Defender who said he would take them and deliver them to the judge personally, the judge had not.

9. They continue to stalk and violate me in every unimaginable way and seem perfectly content with quietly waiting for me to shower as they have quieted down specifically after I wrote the District Court, Secretary of the Department of Homeland Security and the President after many, many similar letters up the Chain of Command. They refuse to leave the bathroom and are using this tactic as a way of gratifying an obvious sick breast fetish and to further terrorize, and painfully humiliate me. If I don't shower I hear comments that I am unclean, nasty. When I do I am trying to be sexy for the men who are watching. When they learned that I had gotten the form to file a complaint on my own behalf in this court, the older woman, whom I have seen before, said good if she does then when can prove that she is competent and expert on the law and that will void out the Rule 11 Hearing. They are relying on DUI charge as they know that a felony would terminate my VA and Social Security benefits. I did not asked for a Rule 11 Hearing and did not know what it was when Ms. Wallace, Mesa Public Defender, requested it.

10. As I watch movies inside my apartment they try to build a case. Of nearly 200 movies I have checked out from the Blockbuster Video Club, a few, ten to be exact, have had sexual assault content which is not described on the back cover. They say "Oh, yea we got you know or whisper look at what she's watching, she wasn't raped" I have no

desire to watch this type of movie under any circumstances and once I learn where the movie is heading I immediately take it out. However, they insist that I am intentionally checking these movies out and make comments for me to hear them inside my apartment. If I listen to a song in my car written by a white and black artist about Hurricane Katrina they say "She angry about America's treatment of blacks." If I am startled during a suspenseful movie they say I am faking a startle reflex which I was diagnosed with long before they came along. If I put on my Bob Marley hat after washing my hair, they say that I am a black radical, anti government and mad about slavery. I have never been discriminated against, my benefits were given after a lengthy investigation and I am proud to be an American. Although I was in the military for a short time, I made Sergeant after a little over 2 and a half years which was to be officially pinned on after reenlistment. I received an Army Commendation Medal my first year, received an Expert Badge on the M16 rifle, received a Good Conduct Medal and was on the personal staff of a Two Star and a Four Star General before having symptoms directly related to the sexual assault, I was told later during therapy years later. I got out one month after reenlisting. I had planned on making the military a career.

11. Why have they been allowed to violate and victimize in such an unconstitutional manner? I have been taking notes and documenting this experience and each and every occurrence which has angered these people. I was told by the VA who said it would be therapeutic to write down what was happening. The people watching me, who sound like the same people from California still but now partnered with Arizona law enforcement, have stepped up their threats of "Yea, she really, really has to go now!" and they sound serious. They have threatened to tamper with my medication. I heard this from the cop who showed me his penis. Saying, "I know how we can get her through her medications." I continue to live with this ongoing persistent harassment, threats and degradation, humiliation which has spilled over into my private life and destroyed friendships, opportunity for advancement through education, along with their using people trying to entrap me and have prevented me from starting new relationships for fear of their intervention which has lead to isolation as old friendships diminish and are destroyed as a result of this. No one believes me and I am laughed at and ridiculed. They have my neighbors watching me with the X-Ray Vision equipment although they know that I am sober for 2 years as of September 10, 2009 yet tell law enforcement in each location I live that I am actively using drug knowing full well this is a lie. They continue to insure that I hear their comments inside my vehicle and in each and every public places I visit. They laugh at me at home as I go about my daily activities and make comments to hurt my feelings. After their threats about tampering with my medication, I went to Wal-Mart and purchased an alarm for my door. While I was there I heard them say "Oh yea, she's scared now." I recently joined Bally's Gym here in Glendale, Arizona and as I changed into my swimsuit inside the locker room a women said to a man that I covered my panties intentionally because I did not want him to see how dirty I am. On another occasion she told him I was not covering my underwear because I wanted him to see that it was clean. When exiting the shower at home, I wrap myself in a towel as many people do, the women comment that I am doing this (walking around in a towel as I prepare what I am going to wear because the men are watching.) If I get out of the bathtub and look in

the mirror as I put on my bra the women comment that I am doing this for the men's pleasure because they are watching. If I want to give myself a breast exam I am reluctant as this is viewed as some sort of sexual gesture as was one day as I stood in the mirror looking at my abdomen and how much it protrudes due to the fibroid tumor. If I put on a nightgown the women say I am wearing it because the men are watching. If I put on a t-shirt they comment "look at what she has own." They women understand how sick the men are about breast and try to show them that I am nasty and a slut in an effort to steer them away from any advantage, they feel I may have because of them watching me naked and their fascination with my breast.

12. In July of 2008, the VA in California approved for me to attend school with the hopes of returning to the job market. The investigator's ruined this opportunity for me as they visited the school and the radio station where I was Interning insinuating that I was possibly a drug dealer, drug user or purchasing on the premises. I had been sober over a year by then and they knew it as they were watching me 24/7.

13. They become angry as I talk about them harshly after their continued, non stop degrading treatment inside my apartment and bathroom and everywhere else. As they offensively proclaimed "We're going to get her" or "God, I can't wait to get her" and "I can't stand her." I was discredited and harassed to the point that I had difficulty sleeping at night, the nightmares returned with a vengeance (twice the guy who tried to sexually seduce with showing me his penis called out my name from the outside of the apartment while I slept inside in the early morning hours. "Renee!" he yelled.) Due to the strain, stress and embarrassment by their insinuations, I dropped out of the school 2 months before graduation.

14. How much longer must I put up with their violation and false accusations? Why are they allowed to violate me in such a manner? Even as I typed this document, I noticed the cursor moving around on its own and I heard them comment about this writing. In California as I would typed letters of complaint to the appropriate senior officials my letters would be highlighted then deleted and I would have to retype them. Recently I wrote the Department of Homeland Security and the President after I learned that the Department of Homeland Security had decided against utilizing a LE Spy Satellite Program for local law enforcement after running a pilot program with the Los Angeles Police Department with Chief Bratton of the Los Angeles Police Department agreeing with the decision. The decision was made on June 23, 2009 and Congresswoman Jane Harmon was involved saying that the program has potential for personal intrusiveness violations without legal guidelines in place. I find it hard to believe that the Feds would violate me so disgustingly and unprofessionally but believe that it has been a joint effort between the Feds and local law enforcement. I wrote the District Court in Los Angeles, California and did receive a call back as the cops listened in on what should have been a private legal conversation as the person tried to direct me on filing this official complaint. I could hear them trying to determine how much me and person I spoke with knew about who exactly they are. After they felt satisfied that we didn't know anything. (Prior to this, the threats had started to sound really serious, angry and vindictive.) They even

said they would get me as I left my apartment one night to return videos just before midnight. Three gang banger looking teenagers appeared glaring at me before I left and were still there when I returned here in Arizona. This happened in Oakland also after they said in the background "Why not have her beaten up." They consistently take credit for these incidences in the background of my cell phone. There is a way, I believe to find out exactly who they are possibly with some information I was provided.

15. For the past 3 years I have not had a moment of privacy for the most intimate activities such as bathing, eating, dressing, sleeping, talking, communicating, the gym locker room, in court, in jail, meetings and conversations with the public defender, etc., during hospital doctor appointments, medical and mental health therapy, at the grocery store, in my car, etc., etc., etc., and have had to listen to angry verbal comments dehumanizing me during each of these activities. Highly technical equipment has been used continuously in the privacy of my apartment under the guise of a legal investigation without regard for me as a human being, as woman, mother, and a United States Citizen. My grief, from the manner in which the investigation was and is conducted and the horrendous manner in which I am treated, violated and objectified with terrorist comments and innuendo, has cause psychological damage I believe beyond repair. I am forever changed and scarred. Even now as I become angry and lash out they laugh and say that they are going to get me I am physically sickened by this unjust treatment which is definitely magnified by their false accusations and continued, relentless disgraceful behavior.

16. What is proper and correct against what is not to be expected of a law enforcement agency expected to behave in an appropriate manner with integrity and professionalism and utilize common sense and standard rules of conduct normal to correct common sense adult behavior? How should I not expect for these individuals to use intelligence and treat me fairly and decently within the frame of guidelines established procedures. These individual have more in common, based on their tactics, with a Peeping Tom, a Stalker or someone who practices voyeurism and terrorism. What must I reasonably expect as United States Citizens from law enforcement if not common sense, ethics, morality, honor and integrity and logic in conducting a proper professional investigation. At what point do we excuse their behavior as ignorance of prevailing rules of conduct in absence of standard operating procedure for use of spy satellite technology. These individual crossed the line knowingly and with blatant disregard for me as a human being. How can they not know that what they are doing is dehumanizing, inappropriate, disgraceful and dishonorable as they are allowed to practice this barbaric behavior unrestrained and with blatant disregard for any professional rules of conduct, consideration of me as a human being as one would normally expect from an organization sworn to uphold the law, honor, protect and serve honorably the laws of this Country.

17. Just recently, as I left meeting with a therapist, at the VA hospital regarding this situation and the Military Sexual Trauma (MST), they said "We are going to get her by her comment of whether the house where the assault took place was owned by a soldier or not." as I walked to my car. I have always been unsure of this moot point as they try and

negate the brutal sexual assault I experience to ownership of the apartment. Why are they allowed to watch me via spy satellite during confidential hospital appointments and appointments regarding legal matters? Today, August 11, 2009, as I spoke with an attorney regarding guidance in completing this form, they said that "That document will never get filed." As I closed out my day and walked into Boston Market for dinner, around 5:00 p.m., two of the California investigators (a female and a male) came in trying to intimidate me. I guess they are trying to back up their threat that my case will not get filed. I am continually living with threats of "Oh yea, she's really got to go now!" Last week I feigned an attempt at mailing the VA a letter regarding this situation. I find it necessary to do this after realizing as of Sunday, August 9, 2009, that they have no intention on stopping the degrading bathroom behavior. I again listened as they scratched the technology around in the wall in the bathroom, although barely audible. They still were there waiting on me to take off my clothes and heard a faint "Damm!" when I did so knowing that I am filing this complaint, they did not care. And after I realized that I had in fact heard them, I called my other cell phone number and talked about them while they listened in the background as they normally do. I heard one of they say, "So she knew" meaning apparently that I can still tell when they are present, then "She's got to go!"

18. August 13, 2009, I am back in Court regarding the DUI proceedings and they have threatened that if I file this form, I am a law expert and this will void out the Rule 11 Hearing. All technical legal matter in this document, with the exception of the complaint itself was copied verbatim from a ACLU case filing of a similar circumstance I found on the internet. I also have been bluffing them insinuating that I am writing a book and gave information to a Ghost Writer before leaving California. This is not true.

19. The First Amendment provides that in relevant part that "Congress shall make no law. . . Abridging the freedom of speech, or of the press."

20. The Fourth Amendment provides that "(t)he right of the people to be secure in their persons, houses, papers, and effects, against unreasonable searches and seizures, shall not be violated, and no Warrants shall but upon probable cause, supported by Oath or affirmation, and particularly describing the place to be searched and the persons or things to be seized."

CAUSES OF ACTION

21. The program violates the plaintiffs privacy rights guaranteed by the First and Fourth Amendment.

22. By seriously compromising the privacy rights of Ms. Mitchell, the Program violates her First and Fourth Amendment Rights.

23. I, Renee Pittman Mitchell respectfully seek a declaration that the Program is being unlawfully administered as applied to me, that an abuse of power has occurred and that rules of conduct and logical humane operating procedures were not enforced thereby

In one of my Blogs on my WordPress blog websites, is the authentic recording captured as a voicemail after a Federal Agent tried to have me committed to the psych ward of the West Los Angeles Veterans Hospital. For Ms. Sarteschi to say that federal agents, or cops, do not tap phone lines and listen in the background, or comment trying to intimidate and bully, or incite fear, is utterly ridiculous. This is a typical tactic.

The voicemail is from a doctor at the VA hospital who a Federal Agent called, at home, on her off day demanding that I be committed for challenging him with awareness of this program's existence. The link is below:

HOW IT WORKS – Official Use of the TRUTH, to Discredit Beamed TORTURE by using the "Delusional Disorder" Set-up Within the Massive COVER-UP

The transcribed voicemail begins with that same Federal Agent calling me the B word, in the background of my phone, before the doctor talks. He was egotistically mad because I asked her to document in my medical records what he had tried.

he says before she starts talking.

Doctor:

"Hi, Renee. This is Dr. Bernstein-Hass calling, returning your call. I got your message about your following up about the addendum. I spoke with someone at the VA about putting an addendum that will relate to just this and I was given information about how to do that and I've written a brief addendum in your chart which is what I was told I was supposed to do or could do I should say. So, there is documentation of the phone call by me and that is **what they told me to write.** So hopefully that answers your question…"

She left the VA shortly after. She told me, he did not ask but demanded she take actions, again on her off day at home.

Again, the mental illness tag has been used for decades and to today to silence awareness of this program.

Renee Pittman *Mind Control Technology* **Associated Blog's Links:**

Activist Post: NSA Whistleblower Reveals Covert Torture Program – Pulitzer Prize Winning, Chris Hedges Interview

Bigger Than Snowden Original WHISTLEBLOWER Website Taken Down as the Truth Continues to Snowball to Millions Nationwide & Globally

Rogue Federal Operatives – Electronic Surveillance, Microwave Harassment, Subjugation Torture & Mind Invasive Technology

Set-up, COINTELPRO Style, A Historic, Typical, Modus Operandi: Surveillance and Orwell's *1984*, It was Not Supposed to a Manual for High-Tech Mass Population Control

The personnel involved with this program are still trying to influence misconstrued information documented in my medical files at the VA hospital that which does not apply to discredit me today.

On June 4, 2023, I check My Healthy Vets online medical information to see exactly what a newly hired doctor documented during a phone appointment from my home.

I don't mind telling you that I was shocked at the outright misrepresentation and likely again federal agent influence.

What is written below was actually part of a complaint and goal to document everything.

A doctor from India wrote the following:

1. Pt stated that she was studying Sociology. She's reading a book currently about technology being micro-chipped into humans' brains, and what would happen if we didn't have it to rely on anymore.

   Pt states she has written 6 books on the government and its technology. States it's about conspiracy theories related to the government, and she states they are online and are quite successful. (Writer went onto websites, and she is listed as author for several books).

   Pt stated that she had a horrific divorce with ex-husband who was a policeman, and **she did a lot of drugs...**

   First of all, that sounds like bragging and more importantly I did not tell her this because it is 100%, unequivocally, without a doubt, absolutely not true.

   This is unless you are a VA employee as she is and many others they have recruited to play along, that can be influenced by have the same Federal Agents around since trying to get another doctor to admit you to the psych ward. To do so today, the obvious game plan is to discredit you today and do so by keeping ancient history alive. Part of their goal is the create deception and have it documented. This is done to coincide with the narrative they have spun to the public, which again they used to justify monstrous human experimentation nationwide.

And, where in God's name did she conjure up microchipping?

Her writing this is absolutely bizarre and I must admit makes her professionally, highly questionable. I have never had, absolutely any interest in microchipping in my body nor have I told any VA doctor, documented in another doctor's report, in 2012 that I said I had a microchip in my knee.

In this recent effort saying that I said microchip is even more ridiculous that her saying that I told her where would we be without it? For the record, many report microchipping but I have never felt I have a microchip anywhere in my body.

The later documented statement makes absolutely no sense revealing this is a fabrication. Human beings don't need microchips and why would anyone wonder where humanity would be without microchips?

I mean really am I supposed to take her seriously after this nonsensical documentation in my health records?

2. Again, why in God's name would I tell her my books, that are based on research, open literature evidence, patents, and historical data including Federal Information Act (FOIA) derived information that are instead about government conspiracy theories?

Frankly I would NEVER say this.

I know from personal experience that the mere mention of any type of Government Conspiracy is grounds for an automatic diagnosis for paranoia and delusions.

3. Frankly what her writing this revealed to me yet again federal agent influence. I never told her I did a lot of drugs because I did not. What I would I look like saying this? The way she put it, it almost sounds like I was bragging. For the record, I find nothing about substance abuse to be proud of and frankly it sickened me. I have always been so much better

Through all of this , for years, this program is designed to influence the target to sabotage themselves, Perhaps in their twisted minds they believe they are unaccountable. Authority figures influencing people is yet another example of the Milgram Experiment that proved people will do basically anything if asked by authority figure, and no matter if it is wrong.

Since many aware are also reporting targeting since childhood, I have always wondered if, this program has access to knowledge in advance of who will grow up to be problems and why many are used for decades of nonconsensual experimentation? Is this the reason for what many report as early childhood human experimentation?

Frankly thousands being used, start connecting the dots that they definitely are being used and have been used for quite some time and experimented on you want to know why you. At the same time the targeting increases, with again and again and again, discrediting high on the list. When I searched for answers if this is even a possibility I stumbled on the article below.

Frankly, as I looked back, I distinctly remember times in my life where I was held in what felt like a surreal state before and after two major influenced events in my life as experimental hoping to destroy my inner self which is ongoing to this day revealed by the major official setup around me.

Once, when this program hoped to destroy and break me by first creating severe depression (brain entrainment) then influencing what they thought would be hopeful addiction, which failed, and the second time after a monitored sexual assault.

The subliminal influence technology this program uses is their horrific ace in the hole believing it will never be exposed or believed that people are doing this to others.

They are 100% using this technology to destroy lives as is happening to me today, show up reinforcing that the person created the horrors in their life which this program factually created by subliminal influence patented technology and official personnel gaslighting the targeted human guinea pigs. The goal is to convince the target that what happened was the result of their own thoughts or the that target was solely to blame which is unbelievably cruel.

This is beyond evil, past, present and today by officiated personnel at the helm of these advancements who are devoid of conscious and meat suits for evil.

An example of subliminal technology, in use for decades, is shown in the following Carnegie Mellon University link, entitled:

Subliminal Mind Control In 1991 in The USA (1991)

"Science had advanced subliminal radio waves to our hearing, our sight and our senses to the time that a person can be 100% mind controlled without the slightest realization. With adding drugs to eliminate a person's willpower, it is hopeless for the mind and our body cannot resist. Our government, educational interests, commercial and religious use it. Sadly, our social and ethnic clubs have joined. Any group of any size can purchase all types of subliminal electronics.

A real problem is the theft and improper use of this equipment. All kinds of dangerous grab ass are occurring. Lawyers, doctors, and psychiatrists are involved. Ultrasonic is the high-pitched frequencies. Infrasonic is the lowest pitch frequencies. both of these are above and below the noise of sound. N.Y. Times, July 5, 1991, tells the only truth.

Our government could cure almost every deaf person today by implanting a radio receiver implant in a dental filling or small disc under the skin on their heads. For normal hearing there are a dozen ways to send subliminal radio waves. At one time most were safe, but now no one is. Billions of dollars are wasted and stolen in the use of this.

As usual, innocent persons have never heard of this. These audio messages are tricking our minds for the mind thinks they is thoughts. Far too many mind control thoughts are negative, mischief making, theft or cause crime. In thirty seconds, mind controllers can ruin anyone...

After years of research, without a shadow of a doubt, the human monsters program likely knew then that I others Targeted today would fulfill our ultimate destiny of exposing this program as they continue trying and trying to destroy everyone using advanced technology today and by any means necessary.

As an avid reader of everything, I remembered reading about the Akashic Record in a metaphysical book when I was young.

WORLD FUTURES
https://doi.org/10.1080/02604027.2019.1703159

Check for updates

Hacking the Akashic Records: The Next Domain for Military Intelligence Operations?

Jeff Levin

Baylor University, Waco, Texas, USA

"In the present paper, material is presented which cautiously reviews the possibility of a post-cyber domain for intelligence operations, founded on the esoteric concept of the Akashic records—a repository of information and sensory/thought impressions "located" in the nonphysical realms akin to Jung's collective unconscious—thus moving quite beyond the pre- sent five-dimension doctrine. A new doctrine, made operational, would draw on human resources that would seem to surpass current consensus definitions of human capabilities, and would interface with (meta-) physical realities that would seem to surpass current consensus definitions of physical reality. An Akashic domain for military intelligence would thus represent a substantial expansion of the concept of battlespace to include a "dimension" that is located, apparently, outside of space—and time—as conventionally understood.

Former DIRNSA, Commander of USCYBERCOM, and Chief of the Central Security Service, ADM Michael S. Rogers, in advocating for a highly trained, innovative, multi-sector Cyber Mission Force, has stated:

Improving security for all and achieving cyber resilience takes a broad and coordinated effort with unprecedented degrees of joint, interagency, coalition, and public-private sector collaboration. We need sustained interaction, exchanges of ideas, and regular exercises that bring the military and civilian cyber communities together across government, industry, and academia to share information, coordinate planning, exercise, and brainstorm together. (Rogers, 2015)

The present paper endorses the same approach, but applied to a hypothetical next domain beyond cyber, the Akashic domain.

The Akashic records are defined by esotericists and mystics as a permanent record of all of the thoughts, feelings, and actions that have ever occurred in the history of the universe (Bacheman, 1973; Gaynor, 1953), stored in *a kind of cosmic memory bank* (Watson, 1991, p. 6) that exists outside of physical reality. This concept originates in the Sanskrit word akasha (*the ether, and has been described as a field accessible psychically and via spiritual practice (Laszlo, 2007). First introduced to the West by 19th-century Theosophists, the concept of an Akashic records has become part of the lingua franca of contemporary New-Age spirituality* (Levin, 2019)…

One thing for certain, this program is an official monstrosity and great effort goes into keeping the morbid truth hidden. When I finally figured who and how, exactly, I filed the following District Civil Rights Complaint in Los Angeles, California. See Reference Link: Wikipedia **Akashic Records for details**

Following is the 2014 United States District Court of California

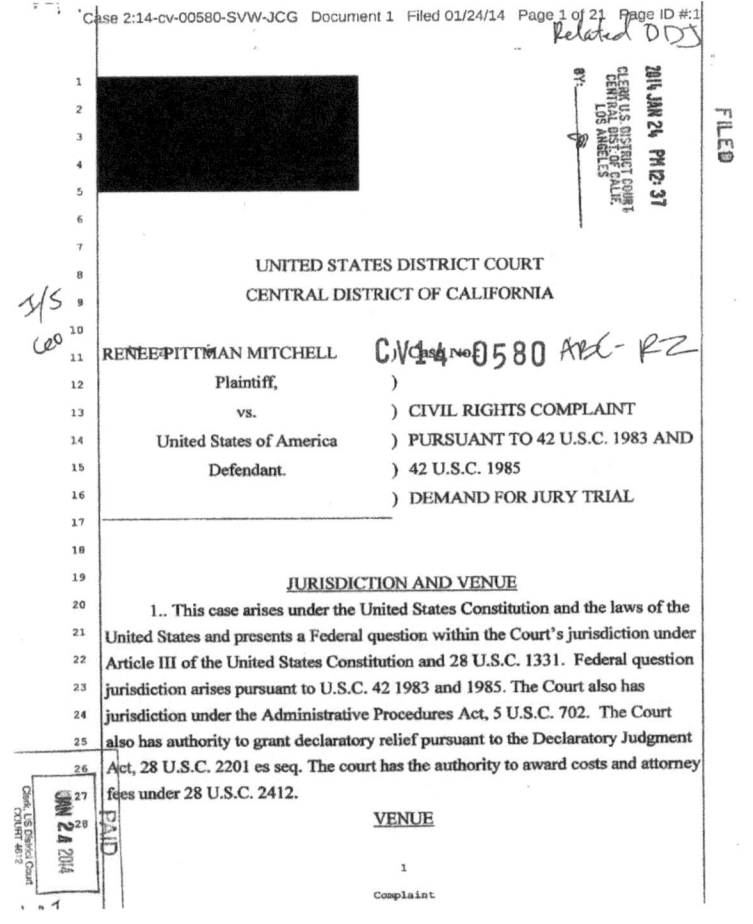

2. Venue is proper in this District under 28 U.S.C. 1391(e), with regard to state law claims, under U.S.C. 1367 because the Defendant is the United States and Plaintiff lives in this district.

PARTIES

3.. RENEE PITTMAN MITCHELL, resides at 3568 Sterling Court, Palmdale, California, 93550, activist/whistleblower is at all times relevant to this matter as a disabled female veteran, deemed 100% service connected, after a lengthy Department of Veterans Affairs (DVA) investigation in 2002, by the United States Department of Veterans Affairs, and later approval of Social Security Disability by the Social Security Administration, (SSA) in 2008, after another, lengthy, competent, and thorough investigation due to physical ailment. RENEE PITTMAN MITCHELL is a United States citizen appearing Pro Se. Ms. Mitchell brings this complaint before the Court for violations of her individual and associational rights under the, 1st and 8th Amendments, to the United States Constitution, resulting from violations of Title 18, U.S.C., Section 241 Conspiracy Against Rights, Title 18, U.S.C., Section 242 Deprivation of Rights Under Color of Law, Title 18, U.S.C., Section 245 Federally Protected Activities and Title 18, U.S.C., Section 249 Matthew Shepard and James Byrd, Jr., Hate Crimes Prevention Act related to a Plaintiff with disability. Plaintiff sues for violations of Civil Rights pursuant to U.S.C. 42 Section 1983 and 1985. All allegations in this complaint are made on information and belief, except to the majority of events in which the Plaintiff is personally involved.

4. UNITED STATES OF AMERICA, resides at Attorney General of the United States, U.S. DOJ, 950 Pennsylvania Avenue, NW, Washington, D.C., 20530-2919, is at all times relevant as the Defendant through Doe(s) consisting of military, law enforcement, and, where applicable, Department of Defense contractor. The true names of the Doe(s) involved are unknown to the Plaintiff working on behalf of the United States who are responsible for the egregious

immoral events in this complaint and proximately caused damages to the Plaintiff. Plaintiff is unable to ascertain the true names of Doe(s) because she was never privy to beforehand machination or Defendant's decisive conduct committed in connection with the United States' overboard, warrantless targeting of her, and as a result of no official notification of action, the names are unknown. Plaintiff will seek leave of court to amend this Complaint to allege such names and capacities as soon as they are ascertained if possible in light of continued monitoring 24 hours a day, 7 days a week to prevent Plaintiff's efforts, by the United States resulting in full knowledge of this action through computer monitoring of Plaintiff's efforts and tampering.

5. The United States of America is *Respondeat Superior*.

STATEMENT OF FACTS

6. In June of 2011, Plaintiff began self-publishing a series of what today are three explosive tell-all books entitled the "Mind Control Technology" books series focusing on a combined effort by United States Air Force personnel, law enforcement, and a group of African American men privy to electronic harassment technology torturing her, in a "slow kill" operation designed to force Plaintiff to stop publication of the books published on Amazon.com through Amazon's subsidiary book publishing company Createspace. This was done repeatedly by accessing Plaintiff's computer and distorting book information, vicious verbal attacks on her phone line, and through tampering with Plaintiff's websites promoting information of the books through knowledge gained by Electronic Surveillance of Plaintiff's personal passwords. The books are published under the pen name Renee Pittman M. Book One in the series is titled "Remote Brain Targeting," Book Two, "You Are Not My Big Brother," and Book Three, "Covert Technological Murder." Book Four "The Diary of an Angry Targeted Individual," continues Plaintiff's Plight and day to day effort of exposure of electromagnetic technology usage and Plaintiff's hope for justice and relief. Book four is set to be

released in early 2014. Plaintiff has also been contacted by a film company regarding said books which has escalated tortuous efforts to stop exposure with ridicule of Defendant that "You can't prove a thing" as Plaintiff attempts this legal action by the African Americans. The book's overall content are that of a very detailed and comprehensive, well-written account of circumstances in Plaintiff's life based on "Electronic Harassment," legalized under the "Exception Clause" of U.S. Code, Title 50, Chapter 32, Section 1520a, authorizing technology testing for the military and law enforcement, without informed consent, to a target testing Electronic Harassment technology for riot and crowd control purposes. The books also include information on DOD Regulation 5240.1.R. This DOD regulation sets forth procedures governing the activities of DOD intelligence components that affect United States persons. Executive Order 12333, "United States Intelligence Activities," which stipulates that certain activities of intelligence components that affect U.S. persons be governed by procedures issued by the agency head and approved by the Attorney General. Specifically, procedures 1 through 10, as well as Appendix A, herein, require approval by the Attorney General. Procedures 11 through 15, while not requiring approval by the Attorney General, contain further guidance to DOD Components in implementing Executive Order 12333 as well as Executive Order 12334, "President's Intelligence Oversight Board," along with details on the Federal Intelligence Surveillance Act (FISA), Electronic Communication Privacy Act (ECPA), PATRIOT Act and the National Security Letter role in warrantless, "Electronic Harassment" efforts of U.S. citizens in the series. The books are also substantiated by unclassified documents, open literature evidence, newspaper articles, television documentaries, testimony of other victims, and detailed personal experiences of Plaintiff regarding unethical electronic harassment technology usage of extremely low frequency (ELF) weapons in various forms, portable, handheld, land, sea and space-based, patented at the United States Patent and Trademark Office. The books give a chronological history of research, testing, and development programs for this technology dating back decades as the foundation for full deployment today of the technology by the military, law enforcement with access by Department of Defense contractors. The effort around Plaintiff appears to be and continues to be unofficial in that Plaintiff was never officially notified and focused specifically on denial of Plaintiff's 1st Amendment Right. The book series detail "Electronic Surveillance" strategies, tactics, and use of physical torture as a form of coercion through bio coded energy weapons deployed from operation / Fusion centers across America to include agency motivated and mobilized community vigilante groups resulting from disinformation campaigns directed at Plaintiff which has erupted into extreme community harassment as an arm of the United States America targeting. The overall objective, consistently, has been to force Plaintiff into stopping promotion of these books, in violation of Plaintiff's 1st Amendment Right of Freedom of Speech, etc.. Physical torture of Plaintiff quickly began to escalate due to book detailing unethical, immoral, and embarrassing in the manuscript phases also as acts specifically by the African American men working on behalf of the United States of America then materialized into their relentless attempts to torture Plaintiff mercilessly into submission through psychological operations and physical irradiation of toxic levels of non-ionizing radiation from their operation centers, and again, also emitted from neighborhood residences parallel to Plaintiff's home to include garages involved in Neighborhood Watch and Community Oriented Policing Programs. Since a crime has not been committed by Plaintiff and Plaintiff has never been charged with anything or arrested, Plaintiff comes before the Court asking for Court intervention against, the vicious, unjust "Electronic Harassment" effort consisting of extreme, psychological and physical torture designed to stifle book marketing and promotion as a violation of her 1st Amendment Right. Her efforts are a plea for her life, due to the "Slow Kill" effect of Electronic Harassment technology being used when focused, relentlessly on a target nonstop.

Court intervention is vital in that this type of official and unofficial targeting which are routinely ruled as inconclusive allowing for the continued gathering of intelligence, surveillance for years, resulting in ongoing physical torture and psychological abuse without end. Court intervention is the only recourse to inhumane tactics and strategies in the effort, and without Court intervention typically could result in Plaintiff "slow kill" death. The effort involves communication equipment, surveillance equipment, extremely low frequency technology, physical force equipment, legalized for testing by aforementioned "riot and crowd control" testing set forth by experts for non-lethal weapons use and application. The foundation for this Civil Complaint is brought before the United States District Court due to efforts of horrendous violation of Plaintiff's 1st and also 8th Amendment Rights in the form of "cruel and unusual punishment" motivated by an intense effort then to deny Plaintiff Freedom of Press, etc., etc., etc., as a Constitutional Right to tell her story, and resulting malicious torture as coercion against the United States of America attempting to stop this right, through acts in direct violation of the 1st, and 8th Amendment Rights of Plaintiff by those employed and/or acting on behalf of the United States of America hereafter known as Doe(s), who began arbitrary targeting Plaintiff which intensified during the manuscript process and continues to intensify after book publications to this day. The United States as the Defendant does not qualify as being impartial or having the ability to be fair under the circumstances, and instead continues inhumane Electronic Harassment. Specific strategic acts evolved designed to enjoin Plaintiff into stopping book promotions by the United States of America and Doe(s) on behalf of the United States of America, seeking leverage against Plaintiff's publicity and exposure efforts through the series of books in the media, radio shows, etc. and her detailing publically her plight. Plaintiff was injured by the actions of the Defendant as a direct result of the Defendants' unconstitutional behavior, tactics, strategies, denying Plaintiff's rights through the use of advanced radio wave, technology focused on Plaintiff resulting in horrific physical pain, emotional distress, also in the form of verbal degradation captured and documented, as an example, in the background of Plaintiff's phone line, by wiretap after the effort failed to coerce Plaintiff. Doe(s) monitored Plaintiff's book marketing and consumer interest on Plaintiff's websites containing substantiating evidence of official patents as a public awareness resource, and various laws approving Electronic Surveillance and technology testing at: http://www.bigbrotherwatchingus.com, which results in repeated mischievous tampering and intentional vindictive acts by Defendant designed to discredit Plaintiff as inept and illiterate as a Defendant as a typical strategic effort to portray Plaintiff as lacking credibility. This was done by Defendant repeatedly accessing Plaintiff's sites, and misspelling and misconstruing words, and by creating incorrect grammar, and punctuation as tampering. The Electronic Harassment centered not only around Doe(s) attempting to stop unwanted exposure of the new technology in use, but also character unbecoming and dishonorable by Doe(s) working in shifts, 24 hours a day, 7 days a week, from government fused operation centers to include sexual exploitation within one of the books. The United States of America and those acting on behalf of the United States of America, military and law enforcement personnel, of which contractors can be used for plausible deniability, in a "fused" effort, systematically, with malice, and vicious intent harassed, and continues to threaten the Plaintiff's life verbally and physically, in violation of the United States Constitution violating U.S.C. 42 of 1983 and 1985 of the Civil Rights Act in violations of, and conspiracy to violate civil rights guaranteed by the substantive and procedural components of the U.S. Constitution. The Defendant is liable for over the top electronic harassment and technological physical assaults, negligence, and intentional infliction of emotional distress, negligent infliction of life threatening technological assaults, and resulting malicious prosecution under Federal and State law and state theory of *Respondeat*

Superior which states that an employer is responsible for actions performed by employees within the course of their employment and the *Equal Protection Clause* which also requires each state to provide equal protection under the law to all people within its jurisdiction. During the course of Defendant's review of Plaintiff's activities and around the clock monitoring of Plaintiff, Defendant and one or more of the Doe(s) conspired to deprive Plaintiff of her, 1st and 8th Amendment Rights by, inter alia, taking the steps herein. Such steps included use of the internet to post an image in Google, placed strategically among Plaintiff's book marketing and advertisements, videos, and book ordering information, websites, to discredit the Plaintiff as a "Schizie" as a typical strategic effort to manage the perception of Plaintiff as the author of these books as lacking credibility due to mental illness. This is a tactic many report man report being used to insure psychological and physical torture continues to escalate as coercion. Plaintiff also has a version of "You Are Not My Big Brother" in which multiple errors materialized after final book publication in June of 2012, which later necessitated, after the fact republishing of the book with corrections in January of 2013. Defendant is privy to this information through, again, monitoring the activities of Plaintiff on Plaintiff's personal computer, phone, etc., through use of the Computer Internet and Protocol Address Verifier (CIPAV) software. The CIPAV is a data gathering tool that is used to track and gather data on individuals. The software operates on a target's computer much like other forms of illegal spyware whereas it is unknown to the target that it has been installed or that it is reporting on their activities. Plaintiff is also the host of an internet Blog Talk Radio Show called "The Targeted Individual's VOICE," (Victory of Issues of Covert Electromagnets) (http://www.blogtalkradio.com/renee00124) in which Plaintiff has interviewed many credible experts in the area of Electronic Harassment and the history of the technology in use today by the Defendant, which includes a Physicist, who is also a best-selling author and retired Naval Intelligence

Officer involved with it in the Navy, a Biologist who has published several studies on microwave technology use used to discredit targets as mentally ill, and a well-known Electronic Harassment expert, who also has appeared on Coast to Coast radio several times speaking on the same content, and who is credited with actually coining the term "Electronic Harassment." Specific efforts by Plaintiff continue to escalate torture and vicious extremely low frequency technological attacks on Plaintiff around the clock from both neighboring locations from a United States of America operation / Fusion center. Plaintiff's publication of electromagnetic technology in use today and the capabilities of this technology and use by the military, law enforcement, and others, scenarios of targeting and harassment, has become a threat to those at the helm of the technology testing program around Plaintiff through exposure. Plaintiff has also been a guest speaker on talk shows, other than her own, regarding the books, their subject matter, which continues to escalate extreme attempts to silence Plaintiff and deny Plaintiff 1st Amendment Rights. Today numerous United States programs are funded under Neighborhood Watch / Community Oriented Policing through the Byrne Formula Competitive Grant. A list of Congressional authorized purpose areas, for the formula grant funding programs for these areas presently include 29 specific enumerated purpose areas. In recent years, Congress amended the list to include such areas as homeland security projects and the Neighborhood "Watch like" projects. What is not publicized is that within these mobilized community efforts are individuals sworn to secrecy, who will likely be hostile witnesses, who are allowed to partake in illegal acts in the form of civilian harassment through portable technology. Plaintiff alleges that this program of Doe(s) is being used, through disinformation in the community to mobilize her neighbor's against her and allow technology to be deployed from their homes and garages. Plaintiff has spoken with several neighbors whose homes have been set up to deploy electromagnetic energy weapons from their garages and rooms in their homes parallel to the plaintiff's home. Although they won't admit involvement, bio-coded directional energy weapons, similar to the Raytheon portable "pain beam" weapon beam has been curtailed briefly after these conversations. However, two houses set up on Plaintiff's block consisting of United States Air Force personnel continue deployment of the powerful "slow kill" directional beam along with nightly attacks from the ceiling downward indicating the operation / Fusion center. Signals intelligence collection of a target within the United States is governed by the Foreign Intelligence Surveillance Act (FISA) and domestically by the Electronic Communication Privacy Act (ECPA) and also Executive Order 12333 & 12334. Signal intelligence is intelligence gathering by interception of signals, whether between people, or whether involving electronic signals not directly used in communication, such as electronic intelligence or a combination of the two. The ability to intercept communication depends on the medium used. The medium can be radio, satellite, microwave, cellular or fiber optics. The role of satellites in point-to-point voice and data communications has largely been supplanted by fiber optics used to track Plaintiff around the clock. Optical fibers are widely used in fiber-optic communications, which permits transmission over longer distances and at higher bandwidths (data rates) than other forms of communication. The Plaintiff is being tracked by computer generated biometric software, and bio-coded energy weapons which map, DNA, gait, iris, and other physical or also by Electroencephalography (EEG) computerized satellite radar. Electro-encephalography is the recording of the prevalent electrical activity around the scalp. Plaintiff seeks relief from horrific Electronic Harassment, in the form of physical torture, psychological harassment, and verbal insults and attempted slow kill murder through legalized use of technology also approved by the following laws: The Foreign Intelligence Surveillance Act of 1978 ("FISA" Public Law 95-511, 92 Stat. 1783, enacted October 25, 1978, 50 U.S.C. ch.36, S. 1566) as an Act of Congress which prescribes procedures for the physical and electronic surveillance and collection of "foreign intelligence information" between "foreign powers" and "agents of foreign powers" (which may include American citizens and permanent residents suspected of being engaged in espionage and violating U.S. law on territory under United States control) of which Plaintiff has never been involved, domestically the Electronic Communication Privacy Act (ECPA.) the ECPA which also has characteristics similar the PATRIOT Act, and to the National Security Letter, again, U.S.C. Title 50, Chapter 32, Section 1520a, DOD Regulation 5240.1.R, Executive Order 12333& 12334, and specific U.S.C., Title 50, Chapter 36, Subchapter 1, Electronic Surveillance Laws. Several efforts have been made to expose and curtail the use of Electronic Harassment technology by United States citizens by literally dozens of targeted individuals with one case, *James Walbert vs. Jerimiah Redford*, in Sedgwick County, Kansas, 2008, in which the court recognized Electronic Harassment and ruled in James Walbert's favor for a restraining order however James Walbert lost attempting to get the order enforced by law enforcement and subsequent lawsuit against the United States of America. Plaintiff realizes difficulties in proving official, documented, technology testing programs, and even enacted laws for operating radio wave technology exists with use of microwave directed energy weapons (DEW) also knows as the Active Denial System to include documented ultrasound weapon usage, such as Long Range Acoustic Device (LRAD) documented to be in full use by law enforcement, such as LAPD, NYPD, and the Sheriff's Department also, today, for example. Unawareness of official use of this technology for riot control has proven to be a strategic beneficial use to those operating in these programs and targeting citizens around the clock without moral oversight and monitoring. However, Patriots such as Jim Guest, (D-Missouri) State Representative, and Dennis Kucinich (D-Ohio) through the Space Preservation Act of October of 2000 tried and whose efforts also substantiate the existence and use of these weapons: The Space Preservation Act excerpt reads: any other unacknowledged or as yet undeveloped means inflicting death or injury on, or damaging or destroying, a person (or the biological life, bodily health, mental health, or physical and economic well-being of a person)…through the use of land-based, sea-based, or space-based systems using radiation, electromagnetic, psychotronic (psychological electronic), sonic, laser, or other energies directed at individual persons or targeted populations for the purpose of information war, mood management, or mind control of such persons or population. As in all legislative acts quoted in this article, the bill pertains to sound, light, or electromagnetic stimulation of the human brain. California has outlawed employers forcing employees to get the Radio Frequency Identification chip (RFID). State Representative, James Guest, (D-Missouri) sought ethical action among his colleagues in a letter dated October 10, 2007 which reads: *Dear Member of the Legislature and Friends: This letter is to ask for your help for the many constituents in our country who are being affected unjustly by electronic weapons torture and covert harassment groups. Serious privacy rights violations and physical injuries have been caused by the activities of these groups and their use of so-called non-lethal weapons on men, women, and even children. I am asking you to play a role in helping these victims and also stopping the massive movement in the use of Verichip and RFID technologies in tracking Americans. Long before Verichip was known we were testing these devices on Americans, many without their knowledge or consent. With the new revelations of the cancer risk besides the privacy and human rights problems with the use of Verichip and RF signals, I am asking for your help in stopping these abuses and aiding those already affected. Your attendance is therefore requested at a conference call regarding these issues on Monday, October 29, at 11 am EST. After a period of brief presentations, we will have a discussion of these issues with the intent of creating a way forward toward solutions*…A few states have enacted laws against various types of extremely low frequency technology and its use such as: Michigan with Public act 257 of 2003 makes it a felony for a person to "manufacture, deliver, possess, transport, place, use, or release a harmful electronic or electromagnetic device for an unlawful purpose." Also made into a felony is the act of causing "an individual to falsely believe that the individual has been exposed to a... harmful electronic or electromagnetic device." Maine with Public law 264, H.P. 868–L.D. 1271 criminalizes the knowing, intentional, and/or reckless use of an electronic weapon on another person, defining an electronic weapon as a portable device or weapon emitting an electrical current, impulse, beam, or wave with disabling effects on a human being. California ban microchip implants when Governor Schwarzenegger signed Senator Joe Simitian's Senate Bill 362 which prohibits employers and others from forcing anyone to have a Radio Frequency Identification (RFID) device implanted under their skin, however, in light of bio-coded, biometric tracking and bio-coded weapon continues to advance as the norm. As Plaintiff continues book promotion and whistleblowing activities, Plaintiff's endures crippling organ damage through intense energy weapon technology focused on her, 24/7 which is documented to result in necrosis of the area of focus similar to a piece of meat cooked in a microwave oven, resulting in heart afflictions and organ damage to name only a few extreme health risks. Plaintiff's torture has resulted in uncharacteristic and abnormal materialization of energy weapon focused directional beams assaults which deteriorate slowly joints of her body documented as uncharacteristic by the medical profession on two occasions. The United States of America in its leadership capacity is sued related to a conspired effort: to deprive Plaintiff of her Civil Rights and engaged in, or agreed to engage in, overt and covert acts described above in furtherance of a conspiracy, and as a result, violated Plaintiff's clear and well established rights under the, 1st, and 8th Amendments of the United States Constitution. Defendant's activity sought and continues to seeks, to chill Plaintiff's efforts through denial of said rights which extends also to deter judicial intervention, through extreme physical and psychological coercion strategies, acts of intimidation, and by creating in essence, an informal system of denial of said rights through malicious Electronic Harassment coercion. The United States of America, is sued in official capacity by engagement in conduct that was either motivated by evil motives, or intent, and involved reckless and callous indifference, to Plaintiff's, 1st and 8th Amendment Rights. Plaintiff was injured as a result of the actions of the Defendant in that denial of Plaintiff's 1st and 8th Amendment Rights were violated by malicious, physical and psychological, Electronic Harassment. Plaintiff suffered emotional distress, pain and suffering, loss of capacity to function normally, and the resulting emotional crippling effect resulting in an inability and God given right to enjoyment of life. The exact amount of Plaintiff's damages, emotional distress, physical torture, psychological abuse, horrendous verbal degradation through the Defendant's effort, and use of patented legalized technology testing is continually being used resulting in "slow kill" operations around Plaintiff. Plaintiff wants the ability to continue to promote her books, and her talk show without threats, harassment and attempts to force her to stop publication and repeated attempts to discredit her effort and torture. When Plaintiff does speak on issues related to her book publishing, as a guest speaker on other talk shows, surrounding the topic of her book, etc., she wants to do so, and is entitled to do so, free from the fear that she will be subjected to electronic harassment, extreme torture, threats and public calumny.

FIRST CLAIM

(Violation of Civil Rights: *Congress shall make no law respecting an establishment of religion, or prohibiting the free exercise thereof; or abridging the freedom of speech, or of the press; or the right of the people peaceably to assemble, and to petition the Government for a redress of grievances*) (1st Amendment/42 U.S.C 1983)

7. Plaintiff realleges paragraphs 1 through 6.

8. By doing the acts described above, in Paragraph 6, Defendant caused or permitted the violation of Plaintiff's well and clear established rights under the 1st Amendment through patterns and practices of over the top electronic harassment, consisting of physical coercion torture, acts of intimidation, verbal degradation / vicious psychological abuse, and misuse of authority, in an effort continuing to escalate lasting and prolonged well over seven years designed to stifle defendant's book promotion and marketing beginning in the manuscript(s) stages, which continues today as extreme duress through technology authorized under Electronic Surveillance laws. Defendant continues relentless physical and psychological attempts to chill Plaintiff's exercise of the 1st Amendment thereby entitling Plaintiff to recover damages pursuant to 42 U.S.C. 1983 and 1985.

SECOND CLAIM
(Violation of civil rights: "cruel and unusual punishment") (8th Amendments/42 U.S.C. 1983)

9. Plaintiff realleges paragraphs 1 through 6.

10. By doing the acts described above in Paragraph 6, the actions of the Defendant described above in this complaint are, vicious, malicious, deliberate, intentional, and embarked upon with full knowledge of, and in conscious disregard of the harm that would be inflicted on Plaintiff in violation of the Eighth Amendment as "cruel and unusual punishment." Defendant could not have believed that this conduct was ethical, moral lawful within the bounds of reasonable discretion. Defendant thus lacks qualified or statutory immunity from suit or liability. As a result of said intentional conduct, Plaintiff is entitled to punitive damages against Defendant, in an amount sufficient to deter Defendant's future tactics unconstitutionally towards Plaintiff and others suffering similar tactics by Color of Law misconduct of Defendant. Defendant continues relentless physical and psychological attempts to coerce Plaintiff's exercise of the 8th Amendment Right thereby entitling Plaintiff to recover damages pursuant to 42

U.S.C. 1983 and 1985.

THIRD CLAIM

(Violation of civil rights: 42 U.S.C. 1983 and 1985)

11. Plaintiff realleges paragraphs 1 through 6.

12. By doing the acts described above in Paragraph 6 Under 42 U.S.C. 1983 and 1985 Defendant under color of statue, ordinance, regulation, subjected Plaintiff to deprivation of rights, privileges, secured by the Constitution and laws name as stated in Paragraph 6 above., restated solely to the 1st and 8th Amendments. The, violation of "freedom of speech, etc., etc., etc." a Constitutional Right of the 1st Amendment to tell her biographical accounts, and the 8th Amendment against "cruel and unusual punishment" in the form of extreme and deadly Electronic Harassment by tactic, strategies, in the Plaintiff's case using systems and devices thereby entitling Plaintiff to recover damages pursuant to 42 U.S.C. 1983 and 1985.

FOURTH CLAIM

(Violation of civil rights: 42 U.S.C. 1983 and 1985)

13. Plaintiff realleges paragraphs 1 through 6.

14. By doing the acts described above in Paragraph 6:

Under violations of 42 U.S.C. 1983 and 1985 in the form of specific violations which are: Deprivation of Federal Civil Rights, Defendant Supervisory Liability, Defective Policy by act of policy maker, Negligence, Intentional infliction of physical, psychological arm and emotional distress, as *Respondeat Superior* entitling Plaintiff to recover damages pursuant to 42 U.S.C. 1983 and 1985.

REQUEST FOR RELIEF

15. Plaintiff is further entitled to a judgment declaring the Defendant's policy of engaging in such practices, strategies, and tactics, systems and devices as unconstitutional.

16. An actual controversy exists between Plaintiff and Defendant, related to respective legal rights and duties. The Plaintiff requests the Court make a declaration of legal rights appropriate.)

17. Plaintiff contends that she has the right to engage in the types of activities in Paragraph 6, continued book publication, promotion and marketing, speaking engagements, on the subject etc., etc., etc., free from being subjected to electronic harassment, threats, physical and psychological torture, while Defendant does not recognized this right after failure to coerce Plaintiff into stopping.

18. Plaintiff has no adequate remedy at law, and unless this Court grants the injunctive and declaratory relief herein requested, Plaintiff will be unable to exercise her rights and will be irreparably damaged for example, by repeated attack by bio-coded energy weapons which focus sporadically on her heart muscle, and skull, and other organs.

19. Without Court intervention, the effort around Plaintiff would surely continue increasing electromagnetic torture, and the resulting slow kill effect of being targeted around the clock is inevitable by numerous studies by experts by those operating around the clock, working in shifts, through employment of the United States of America.

20. Plaintiff is ethically and morally entitled to Injunction and Declaratory judicial intervention as relief in order to prevent Defendant from continued attempts to chill Plaintiff's efforts through an extensive and overboard extreme effort through compensatory, punitive, including general and special damages including general according to proof, in an amount to be determined at trial and the grant of other relief as the Court deems appropriate.

DEMAND FOR A JURY TRIAL

Plaintiff requests a jury trial. A jury trial is vitally necessary to prompt hostile witnesses, sworn to secrecy, in a herd mentality, under the guise of National Security secrecy, having full knowledge of Plaintiff's claims and technology deployed from residences in Plaintiff's neighborhood and from neighbor's home which would further substantiate this complaint.

12/13/2013
DATED: *Renee P. Mitchell*
                    RENEE PITTMAN MITCHELL, Pro Se

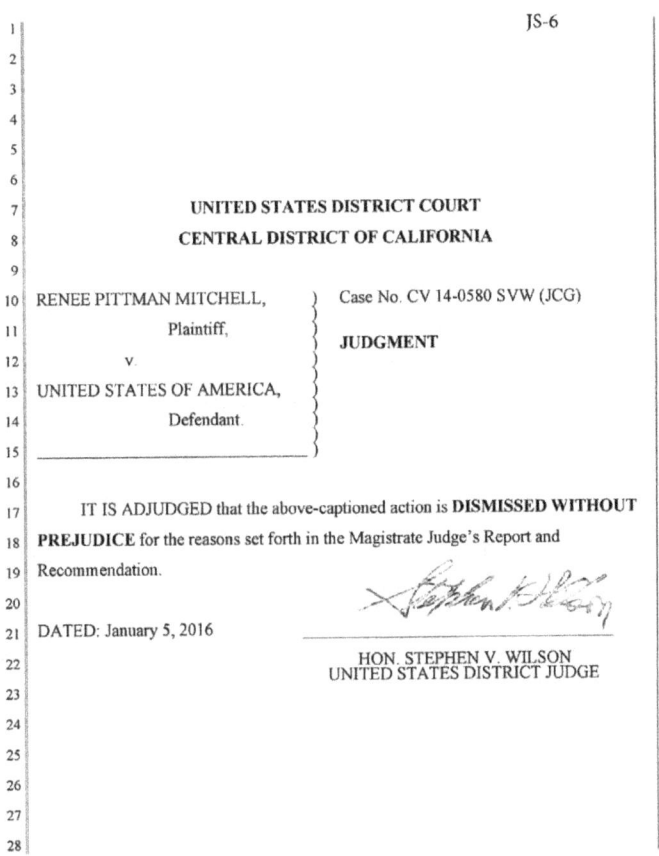

When a case is dismissed with prejudice it is dismissed permanently. A case that is dismissed without prejudice is only dismissed temporarily. This temporary dismissal means that the plaintiff is allowed to re-file charges, alter the claim, or bring the case to another court. Meaning there is future hope with the right and additional substantiating information. Many of these cases are also dismissed With Prejudice for Lack of Subject Matter Jurisdiction.

Again, *Without Prejudice* is a huge step forward. It means that the Court found what your case and what is being reported as feasible and also that there is Jurisdiction.

Lack of Subject Matter Jurisdiction

Subject-matter jurisdiction is the requirement that a given court have power to hear the specific kind of claim that is brought to that court. While litigating parties may waive personal jurisdiction, they cannot waive subject-matter jurisdiction.

This confirms that this is a Federal issue.

CHAPTER 9

The Mechanics of Detecting Voice to Skull (V2K) Official Beamed Schizophrenia

Automatic Machine for Identifying Victim of Abuse Voice to Skull and Remote Neural Monitoring Technology and for Identifying Remote Attacker or Operator Using Device of Voice to Skull and Remote Neural Monitoring

Us Patent 20200275874a1

[Excerpt]

Aug 19, 2020

Automatic machine can both identify victim and remote operator using device of V2K and Remote Neural Monitoring. The automatic machine using databases consisting seven catalogs contents. To launching a retinal image signal attacking remote operator, ask a tested person select and watch one image or video with some words attacking remote operator only causing remote operator's psychological response, but doesn't cause tested person's psychological response, then answers question on a touchscreen according V2K feedback signals by pressing buttons YES or NO, the automatic identifying system will show CORRECT or WRONG with every pressing. This function can guide the tested person to get final accurate identification quickly. Meanwhile printing and storing system can print certificate to victim and documents of all kinds status of remote operator, store them in an information database in computer; provide an access to users in network for updating information.

Description

Cross Reference Of Related Application

This is a Continuation-In-Part application of a non-provisional application Ser. No. 16/558,040, filing date Aug. 31, 2019, which claims the benefit of U.S. provisional patent application No. 62/812,915, filing date Mar. 1, 2019. The content of these specifications, including any intervening amendments thereto, are incorporated by reference herein in their entirety.

Background Of The Present Invention

Field of Invention

The present invention relates to a kind of automatic machine can both identify who is victim of abuse voice to skull and remote neural monitoring technology and identify who is remote operator using device of voice to skull and remote neural monitoring. Then produce a certificate to an identified victim for medical department using, meanwhile produce some documents about remote attacker or operator's status submitting to department of justice for reference.

Description of Related Arts

Origin of the problem: from 2008 to 2018, there were large number of victims in 22 provinces in China, Shanghai and Beijing municipalities have large number of collective protests against RNM and V2K technology abuse. This is also a part of operational level of war or called unrestricted war which China has somehow partly launched to attack US and other countries now or in no long future.

From psychiatric taxonomy analysis, only rare case of hysteria likely to occur in group under special situation. These same groups of large-scale protester's peaceful and rational activities year and year in China are not hysteria patients obviously.

In addition, on Oct. 1, 2015, representatives from 17 countries in Berlin, Germany, hold a Covered Harassment Conference. The covered harassment is same thing with voice to skull & remote neural monitoring technology.

After a lot of times contacting with these victims in China mainland WeChat, emails and telephones, these victims' thoughts were read and harassed at same time by some unknown Chinese institutions abusing these technologies. Meanwhile all victims know their optic neural image signals, or call retinal cortex image signals and auditory neural signals in their cerebral cortex were monitored remotely.

According to Zhong Zhiyong, the organizer and leader of anti-V2k &RNM torture in 22 provinces and cities in mainland China (Zhong Zhiyong's affairs were also reported by the media <u>The Epoch Times</u> and he and other victims' representatives also received interview with *New Tang Dynasty TV*). He told me directly by phone and WeChat almost all victims have the same experience: the Remote Neural Monitoring and V2K technology can read victim's auditory signals on cortex and victim's retinal cortex image signals, harass all victims with V2K signals. Including two Chinese victim has been a lawful permanent resident in United States have same experience: their retinal cortex image signals and auditory neural signals in their cerebral cortex were monitored remotely. They are all suffering non-stop V2K harassment.

Another Chinese leader anti-abuse technology of V2K & RNM, Mr. Yao Dou-jie live in Shenzhen China who collect large information also confirmed same cases above.

Jun. 21, 2020, United Nations Human Rights Office of the High Commissioner call for input to a report: Psychosocial dynamics conducive to torture and ill-treatment, decide to investigate and ban electromagnetic Torture, Harassment techniques torture to individuals. This is first time UN official recognition the existence the abuse V2K and Remote Neural Monitoring Technology.

What is Remote Neural Monitoring Technology? the following US patents describe them clearly: These technology can read human neural signals like human optic neural signals, or call retinal cortex image signals and auditory neural signals in their cerebral cortex to produce harassment to a victim. Because these US patents have external link to European patent database, so all US and European engineers and technical personnel acknowledge the existence of these technologies.

Generally three to five operators (ROD) of V2k & RNM devices in group in turn concentrate one victim via satellite positioning technology to realize remote harassment, the technology is mainly based on these US technological patents and China patents which describe the technology of V2K & RNM are extracting the brain's characteristic wave frequency and their optic neural signals, or call retinal cortex image signals and auditory neural signals in their cerebral cortex first (like fingerprints, each person's brain characteristic wave frequency is different), then codes the brain characteristic wave frequency and these neural signals by computer software technology and remotely locks the frequency and monitors the victim's brain wave frequency and these neural signals using computer software technology and satellite positioning technology.

The bases of these V2K technology are microwave auditory hearing effect or call Frey effect and neural coding technology. The microwave auditory hearing effect is a well-known physical phenomenon and it does not a sound wave shaking eardrum in ear, but directly effects on human auditory cortex with like KAKAKA sound. V2K technology is a kind of improving microwave auditory hearing effect with human language which only a victim can feel, the persons around the victim never hear or feel.

In such way, three to five operators (ROD) using devices of V2K & RNM in groups perform active psychological attacks on victim's brain with some negative emotions according these victim's auditory signal and, retinal cortex image signals shown on the their RNM device. But the effect of this kind of psychological reactions or responses can be two-ways or two directions between victim and operators (RODs) using devices of V2K & RNM.

In past of years, FBI's investigation on some cases related similar technology attacking.

This is not a scientific fiction, because the following US patents and China patents tell us it is true and existing technologies: US patents: All following US patents describe technologies can monitor auditory signals and retinal signals in cerebral cortex remotely.

1.1. U.S. Pat. No. 3,951,134A: Apparatus and method for remotely monitoring and altering brain waves.

   Inventor: Robert G. Malech, Date of publication: 1976-04-20 publication

   The technology describes such technology can monitor human brain wave and all Kinds neutral signals like auditory signal and retinal signal remotely. Meanwhile cause V2K signal feedback like microwave hearing effect.

   Apparatus for and method of sensing brain waves at a position remote from a .... The demodulated waveform also can be used to produce a compensating signal .... In addition to passively monitoring his brain waves, the subject's neurological signals.

1.2. U.S. Pat. No. 6,470,214B1 Method and device for implementing the radio frequency hearing effect, US Air Force.

1.3. U.S. Pat. No. 7,222,961B2 Method for detecting a functional signal in retinal images.

1.4. U.S. Pat. No. 6,011,991A: Communication system and method including brain wave analysis and/or use of brain activity.

Inventor: Aris Mardirossian; Date of Publication: 2000-01-04

The technology describes how communicate human brain thoughts and cause language like effect remotely via satellite.

A system and method for enabling human beings to communicate by way of their monitored brain activity. The brain activity of an individual is monitored and transmitted to a remote location (e.g. by satellite). At the remote location, the monitored brain activity is compared with pre-recorded normalized brain activity curves, wave forms, or patterns to determine if a match or substantial match is found. If such a match is found, then the computer at the remote location determines that the individual was attempting to communicate the word, phrase, or thought corresponding to the matched stored normalized signal.

1.5. WO2014066598A1, WIPO (PCT): Image retinal intrinsic optical signals Remote neural modulation brain stimulation and feedback control. Inventor: Laura Tyler PERRYMAN, Chad ANDRESEN, Patrick LARSON, Graham GREENE. Date of publication: 2014-09-18.

The method, system and apparatus is presented for a wireless neural modulation feedback control system as it relates to an implantable medical device comprised of a radio frequency (RF) receiver circuit, one or more dipole or patch antenna(s), one or more electrode leads connected to at least one dipole or patch antenna(s), and at least one microelectronic neural modulation circuit, and an external or internally implanted RF device to neutrally modulate brain tissue in order to treat medical conditions that can be mediated by neuronal activation or inhibition, such as Parkinson's, Alzheimer's, epilepsy, other motor or mood based disorders, and/or pain. The implantable receiver captures energy radiated by the RF transmitter unit and converts this energy to an electrical waveform by the implanted neural modulation circuit to deliver energy that can be utilized by the attached electrode pads in order to activate targeted neurons in the brain.

1.6. U.S. Pat. No. 8,738,162: Clustering of recorded patient neurological activity to determine length of a neurological event, Inventor: Mark G. Frei, Ivan Osorio, Nina M. Graves, Scott F. Schaffner, Mark T. Rise, Jonathon E. Giftakis, David L. Carlson. Date of Publication: 2014-05-27.

Apparatus and method detect a detection cluster that is associated with a neurological event, such as a seizure, of a nervous system disorder and update therapy parameters that are associated with a treatment therapy. The occurrence of the detection cluster is detected when the maximal ratio exceeds an intensity threshold. If the maximal ratio drops below the intensity threshold for a time interval that is less than a time threshold and subsequently rises above the intensity threshold, the subsequent time duration is considered as being associated with the detection cluster rather than being associated with a different detection cluster. Consequently, treatment of the nervous system disorder during the corresponding time period is in accordance with one detection cluster. Treatment therapy may be provided by providing electrical stimulation, drug infusion or a combination. Therapy parameters may be updated for each mth successive group of applications of the treatment therapy or for each nth detection cluster.

1.7. US20080146960A1, Headset For A Wireless Neural Data Acquisition System, Inventor: Pedro Irazoqui-Pastor, James C. Morizio, Vinson L. Go, Jack D. Parmentire Date of Publication: 2009-12-29.

A head stage for a Neural Data Acquisition System is shown and described. In one embodiment, the head stage includes at least one Input Pre-amplifier, and a multiplexer (MUX) for multiplexing at least one channel. In one embodiment, the input filter of the Input Pre-amplifier is tuned by adjusting the gate voltage of a transistor operating in sub-threshold mode.

1.8. U.S. Pat. No. 4,858,612A Hearing Device, Inventor: Philip L. Stocklin; Date of Publication: 1989-08-22, A method and apparatus for simulation of hearing in mammals by introduction of a plurality of microwaves into the region of the auditory cortex is shown and described. A microphone is used to transform sound signals into electrical signals which are in turn analyzed and processed to provide controls for generating a plurality of microwave signals at different frequencies. The multifrequency microwaves are then applied to the brain in the region of the auditory cortex. By this method sounds are perceived by the ma

1.9. CN 2008202247769.1 China use the tech both in peaceful time and war time.

Publicly published books: All these public books describe some technologies can monitor neural signals like auditory signals and retinal signals in cerebral cortex remotely.

Chinese military research book and public issued books:

2.0. <Control Brain Technology Development and Military Application Prediction and Research> Date of publication November, 2016, Total pages 166 — Doctoral Paper of the Third Military University of China Author: Luo Xue, Principle Professors: Gou Jiwei and Liu Cangli.

The doctoral paper shows Chinese military and government institute develop this technology vigorously which is used in Operational Level War to attack our USA and other countries.

2.1. <Neural Monitoring, The Prevention of Intraoperative Injury>, ISBN 978-1-4612-0491-6 Authors: Salzman, Steven K., eBook $219.00

2.2. **<Remote Brain Targeting: A Compilation of Historical Data and Information from Various Sources>,** by Renee Pittman, eBay:$15.89, Sold by greatbookprices (115117) 98.9% Positive feedback. Also sold on by other major book online sites and now a seven-book series.

2.3. <Surviving and Thriving as a Targeted Individual: How to Beat Covert Surveillance, Gang Stalking, and Harassment>, eBay: US$19.27 ISBN 10: 1549542931/ISBN 13: 9781549542930

Now we have confirmed these technologies existing with US patents and China patents, as well as publicly issued books.

Till now, there is not any method to identify both a victims of technology abusing and those remote operators (RODs) using devices of V2K and RNM around world. These abuse has been developed global and transnational.

Summary Of The Present Invention

This invention uses programmable logical controller (PLC) as main comportment, because PLC is widely used in automatic control areas which input all kind signals to be processed by a special CPU inside of PLC. The invention also input all kinds of electronic signals from V2K feedback signals and retinal image signals transformation to achieve identifying goals.

First the invention is used to identify a victim of abuse V2K and RNM technology, the victim receives coded physical electromagnetic and microwave wave radio frequency harassment called V2K harassment feedback signal, V2K's physical base is microwave hearing effect, or called Frey effect, it is an ordinary physical phenomena, Here it needs to be emphasized the V2K signal is generated by voice to skull (V2K) technological device which is described by above US patents, there is not human factor.

Second why the reason the remote operator can harass the victim is the victim's brain wave frequency was taken in very near distance first by someone and then the brain wave frequency was coded and decoded into some software then using satellite communication technology to harass victim remotely and take victim's neural signal such as retinal image signal and auditory cortex signal, like a radio station receive and send shortwave information with fixed frequency. So, the invention uses of the remote device taking and showing victim's retinal image signal, with pre-designed images and videos with some text words to display on the remote device of V2K and RNM, deliberately causing the remote operator feedback signals of V2K and RNM V2K. Here also needs to be emphasized the retinal image signal is collected by remote neural monitoring (RNM) technological device which is described by above US patents, there is not human factor.

Because the ROD wants to harass a victim, the ROD must use of his or her negative words with V2K technology-based microwave hearing effect to harass the victim via device of V2K and RNM and via satellite communication and obtain victim's emotion reaction from these device, so in this case has a bidirectional and interactive process.

The invention first time uses of such process to launch multiple attacking to ROD in the opposite direction using victim's retinal image signal with images or videos with some text words in projecting database from victim's side (TP's side) to cause response or reaction from ROD. Once ROD makes some responses or reactions which the victim (TP) cannot make, the programmable logical controller (PLC) will make judgement and identification quickly. How to say, the art of war is called to turn somebody' trick against himself or to beat somebody at his own game, because the automatic machine in this invention can change the passive harassment to victim into revealing the identity of victim and further more revealing all kinds of status of ROD himself/herself using device of V2K and RNM, although the ROD stays at a remote place.

Third, the special algorithm running in the CPU in a programmable logical controller (PLC) uses the automatic control curve to identify the tested person is a victim, then can identify the all status of a remote operator using device of V2K and RNM.

Summarize the above three steps, the invention is the world first using Automatic Control Technology — Programmable Logical Controller (PLC) to expose the huge amount abuse of V2K and RNM technology, it is only one countermeasure between Automatic Control Technology and V2K & RNM Technology at present, it is not a battle between human beings.

Different from human brain, PLC does not consider mistakes of electronic signals which a human being inputs by pressing buttons on a touchscreen by a person, because the special software in PLC choose a fluctuating trend which approaches correct goals finally, whereas human brain always considers choosing which one is correct or which one is wrong only, so this job cannot be instead of human brain. The invention does not exist mind process.

PLC is different from ordinary computer, PLC have special CPU which can scan 1000 instructions within 100 ms in average, all kind of Timer (T0-T255) which are convenient for timing any length of time, all kinds of variable memories such as VB store Bit, VW store Word, WD store double words, all kinds of bit memory M 0.0-M 31.7, special memory SM, all kinds of Counters (C0-C255) and high speed counter HC0-

HC5 and so on, so PLC can completes task which ordinary computer cannot complete.

Furthermore, the invention has a very advanced automatic guide system, when a person chooses thousand pictures or videos from projector's classified database and press answer buttons on a touchscreen, the special PLC algorithm will show green light CORRECT or red-light WRONG in the touchscreen within 0.01 second, so the guide system will let the tested person (TP) reach the final identification goal very fast.

At last the automatic machine can automatically print a certificate of victim to a person to identify him/her is a victim which can provide reference to medical department. Meanwhile, the automatic machine can also automatically print some document about remote attacker's or remote operator's all kinds of status which can submit to department of justice for reference.

The invention is a unique combination of existing technologies in the world, using a unique software which the inventor programs, generates a unique effect in the world in a field where on one has ever involved, so no doubt, it is a real invention.

An object of the present invention is to provide an automatic machine can identify both the victims of technology abusing and those remote operator's status using devices of V2K and RNM around world. This is the first set invention of identification combined with polygraph technology, projecting image technology, special automatic technology (PLC) and human-machine interface technology around world which can automatically generate identification results, except checking polygraph meter with human, all other parts of the system achieves automation.

The principle of the invention utilizes the remote operator can read retinal image signals which later transferred into image on a remote device which operator is using according these above US patents description, so the operator can see what a victim sees via

Remote Neural Monitoring technology. The invention asks a victim to watch or only a glance some images or videos with some classified psychological attacking words against remote operator to send these retinal image signals to a remote operator and attack a remote operator directly to cause the remote operator classified psychological response which the victim can feel or sense according V2K feedback signal technology describing in these above US patents description. Then asking the victim to answer designed question on a touch screen with pressing buttons YES or NO,

American eye doctor with US PhD Ms. Diana Chao, M.D. explain very clearly that retinal image signal formed by some lights into retina is not mind process.

When a person glance or watch on something forming a retinal image signal only in a microsecond (equal to one millionth of a second) then transfers to remote device of V2K and RNM with the speed of electromagnetic wave transmission, these is not any mind process in the procedure.

Every image and video from projector database contain some per-designs text words to attack remote operators using device of V2K & RNM technology, so the victim's (TP's) retinal image with such text words only cause those remote operator's psychology reaction, but cannot cause TP's psychology reaction.

So, the system is a kind of automatic machine inputting raw materials and outputting products as following simple diagram presentation, the retinal image signal forming like camera lens imaging function, so there is not any mind processing. Refer FIG. 11 *i*.

Abbreviation in the Invention

V2K: Voice to Skull

RNM: remote neural monitoring technology

ROD: remote operators using devices of V2K and RNM technology

ISA: an identification system administrator

TP: a tested person who originally claims him/her as a victim

PLC: programmable logical controller

PM: polygraph meter (computerized digital)

TS: touch screen

Six Steps for Identification:

Step 1. Ask a TP who claims himself/herself as a victim to use of RODs reading his/her retinal cortex image signals and auditory neural signals in their cerebral cortex being monitored remotely to watch some images or videos with some words in databases in a projector to send retinal cortex image signals to some remote operator using devices of V2K and RNM, (an American eye doctor with US PhD Ms. Diana

Chao, M.D. explain very clearly that retinal image signal formed by some lights into retina is not mind process). This phase may take about five seconds.

Step 2. Then these retinal image signals will be shown on those devices as image with words only attacking ROD will cause those ROD's psychological response and at same time the TP can feel or sense those ROD's psychological responses according V2K feedback signal technology on TP's auditory cortex. The technology comes from US patents and China patents above and the base is microwave hearing effect or call Frey effect. This phase may take three seconds.

Step 3. Ask the TP to answer pre-designed questions which shows a touch screen by pressing buttons YES or NO on a touch screen according which psychological response from RODs. The YES or NO button on a touchscreen via M (Bit memory) communicating with M (Bit memory) in a PLC, via variable memory VW make a judgement and response within 0.01 second via M (Bit memory) to show green light CORRECT or red-light WRONG on the touchscreen. The timer (T) in PLC keeps these showings CORRECT or WRONG on for two sends on the touchscreen. Then the TP see the small green light CORRECT or small red-light WRONG on the touchscreen. The base of technologies come from CPU in a programmable logical controller make judgement with automatic control algorithm and basic human-machine interface asking person to make human-machine interaction. So, from phase 1 to phase 3 only takes 10 seconds.

When step 4 finish, return to step 1 into a circulation with a high speed advanced automatic identifying guide system responds the tested person (TP) pressing YES or NO button only within 10 ms indicating CORRECT with small green light or WRONG with small red light. According automatic control decay curve (PID control regulate curve to prove the identification is accurate via a number of times of successful identification being divided by the number times of failure identification with sending multiple times of retinal cortex image signals to RODs and getting multiple time of psychological responses from RODs. Then the tested person (TP) can choose next image or next video in projecting database. So, the automatic identifying machine can guide tested person (TP) quickly to reach identifying result and nearly does not need tested person (TP) to make some judgment, more than 90% judgments make by the automatic identify machine. As the error curve approaches the X axis gradually. The base of theory come from automatic control theory.

There are a, b, c, d, e, f and g seven databases total, complete one of seven databases taking about 30 times to 50 times reverse attacking to ROD, one time of attacking only takes 10 seconds, so the invention can complete one database within 300 seconds to 500 seconds to reach an identifying result.

Step 5. When these identification finish database a, b and e successfully, the TP has been identified as a victim of abuse of V2K and RNM technology, the automatic machine will show the TP is a victim on the touchscreen and print a certificate of victim automatically.

Step 6. When TP finishes one of seven classified picture or video (a,b,c,d,e,f,g) projecting database identification, ISA connects PM to TP to confirm TP answering all questions on the touchscreen is honest. In case of TP complete this phase is not honest, ISA had to ask TP to repeat this identification phase.

Step 7. The information of victim and information of ROD will be stored into a victim database and a ROD database in standardized format; these information in the database can be real time update by a content server; and there is remote access to users over a network and transmitting the message to users over the computer network when information update.

However, the method only asks the TP to choose some contents in some database to send retinal cortex image signals to RODs when TP see some images or videos with some words in some database in a projector, in this phase, there is not TP's mind process. The next phase, TP answers very simple fixed questions on touch screen according the TP feels or senses which psychological response from RODs according V2K technology signal feedback on TP's auditory cortex, meanwhile the automatic guide system shows CORRECT with small green light or WRONG with small red light on the touchscreen on every pressing YES or NO button, here is nearly not TP's mind process also, because microwave hearing effect or Frey effect does not belong mind process.

According theory of psychoanalysis: the psychological response between an active psychological attacker and a psychological attacked person passively have completely different characteristics, so these different psychological responses caused by selective classified psychological attacks can be used both to identify a victim (TP) and those remote operators using device (RODs) of RNM and V2K via satellite, even the victim (TP) is in USA, but the remote operator is in China.

Patent History

Publication number: 20200390360

Type: Application

Filed: Aug 19, 2020

Publication Date: Dec 17, 2020

Inventor: Da LI (West Covina, CA)

Application Number: 16/997,656

Classifications

    International Classification: A61B 5/0484 (20060101); G06F 3/0488 (20060101); G06F 3/14 (20060101); A61B 5/00 (20060101); G16H 40/67 (20060101); G06F 16/44 (20060101); G16H 20/70 (20060101);

    The cochlea is very sensitive – normal conversation displaces the eardrum by about 10-10 m, which, after transformation by the ossicles, is sufficient to stimulate the cochlea

    Patent no. US 6587729; 2003 Apparatus for audibly communicating speech using the radio frequency hearing effect

    Patent no. US 4877027:1989 Hearing system

    Stephanishen P R (1999) Acoustic Bessel Bullets. *Jnl Snd Vibr* 222(1), 115-143

    Leventhall G (2004) Big Noise in Baghdad, *Noise and Vibration Worldwide*, June 2004, 27 -30

CHAPTER 10

If a Wheel Squeaks Loud Enough it Will Get Oiled

In one of the most uplifting passages of Scripture, we have this reassurance:

He who dwells in the shelter of the Most-High will abide in the shadow of the Almighty (Psalm 91:1, ESV). To dwell in the shadow of the Almighty is to live under the promise of God's protection.

Throughout history, numerous experiments have been performed on human test subjects in the United States on people that have been deemed expendable and many studies confirmed to be inhumane and unethical. The most horrid are experimentation that are performed without human knowledge or informed consent, and in this category unethical experimentation continues using those who are oblivious. Mass experimentation exposing humans to chemical and biological weapons fit the category and widely reported use of an array of weaponization of the Electromagnetic Spectrum

When media frenzy focused on beamed energy weapons, and the resulting condition, termed *Havana Syndrome*, that intended to disrupt or damage the victims' brains, an investigation of this matter reported that the most likely source of the series of mysterious beamed attacks were electromagnetic frequencies, that singled out U.S. personnel assigned to the U.S. Embassy in Cuba in 2017.

Three experts in this field were contacted by the State Department to investigate 25 cases of government employees working in Cuba who complained of suddenly hearing noises, pressure in their ears, followed by symptoms such as headaches, vertigo, dizziness and loss of cognitive functions.

Neuro-weapons are an emerging, but little understood threat that is compounded by the ease of their deployment, and being undetectable to the human eye as invisible beams.

The team consisted of Dr. James Giordano, a professor at the departments of neurology and biochemistry at Georgetown University Medical Center and chief of the Neuro-ethics Studies Program at the Pellegrino Center for Clinical Bioethics.

Neuro-weapons can take the form of biological agents, chemical weapons — and in the Havana case — directed energy, and possibly a combination of different methods, Giordano said. The effects can be latent, durable, highly disruptive and attribution is difficult. They are not weapons of mass destruction, but they have the potential to disrupt everything from individual cells in a body, to societies and geopolitics. Giordano further substantiated his expertise as the authored of a 2014 book, *Neurotechnology in National Security and Defense: Practical Considerations, Neuroethical Concerns*.

Similar to what U.S. citizens are reporting, the attacks occurred communities established for embassy personnel and inside the victims' homes, with the exception of two cases that took place in a hotel known to accommodate diplomats and other U.S. personnel.

Dr. Michael Hoffer, is the director of the Vestibular and Balance Program at the University of Miami's Department of Otolaryngology. Hoffer was the first physician to examine the victims on site in Cuba, and later in Miami

*I thought they were being **targeted**. I still believe they were being targeted*, he said. Those he examined said they were sitting in their homes, or the hotel, when they suddenly felt the symptoms: a loud noise exposure, a pressure sensation, ear pain, then ringing in the ears and dizziness. A day later there were some cognitive deficits reported.

They would actually say that if they moved in their domicile, the beam – as it were – would follow them. Except the minute they opened the front door, it went off, Hoffer said.

Hoffer continued his studies and ultimately carried out a series of preliminary tests on 100 personnel assigned to the embassy, as well as 10 people who were in the same residences of the victims at the time of the attack, but in other rooms. No symptoms were found in the 100 other employees which speaks to the weapon's systems ability to isolate a specific target in a room of people.

Other than speaking to one another about the symptoms, the victims were not influenced by newspaper accounts of the impact and character of these types of weapons. In addition, Hoffer arrived almost immediately after the attacks when the target's memories of what happened were fresh, and more importantly none of them had received any medical treatment which further authenticated Hoffer's tests. He also had the opportunity to view the location where the incidents occurred and would later state *I literally looked and walked the areas where they were hit.*

Coincidentally, most of the targets worked in a particular area of the embassy. Hoffer concluded, those at the helm of the weapon system are experts and that *They were really good at targeting,* meaning the targets were hit with pinpoint accuracy for a desired effect. The homes of the all of the targets were clustered in a neighborhood designed for diplomats.

Giordano, later confirmed that directed energy weapons can cause specific injuries by creating cavitation, or air pockets, in fluids near the inner ear. The shape of the inner ear amplifies sounds and waveforms to cause the effect. The process creates air bubbles, or cavities, that eventually burst. This is because, near the inner ear are two pathways carrying blood to the brain — the cochlear aqueduct and the vestibular aqueduct. The bubbles can travel rapidly up the aqueducts into the brain where they can function as a stroke,

The third member of the team was Dr. Carey Balaam, a professor of otolaryngology, bioengineering, neurobiology and communication and science disorders, at the University of Pittsburgh.

Balaban, expertise was with the Office of Naval Research. He was familiar with a body of scientific research, mostly conducted by ONR from the early 1960s to the late 1980s, on directed energy weapons and their effects on the inner ear. The research sought to identify the human head's susceptibility to directed energy exposures. The papers helped form part of his conclusions. *The likelihood of multiple sources of energy of acoustic and radio frequencies ... is there, we just don't know what it was,* he said

Although the team could not conclude exactly what method used, however, they were able to narrowed it down to several possibilities: drugs alone was deemed unlikely. Ultrasonic (acoustic) exposures they judged very possible and likely. Electromagnetic pulsing was also rated very possible and probable. Microwave energy was deemed possible, but unlikely. Lastly consideration was given to the possibility that the attackers used a combination of a drug that was activated by the directed energy weapon.

The case has drawn a great deal of attention within the Pentagon and at Special Operations Command, he said. The information was recently declassified and the three have given briefings to SOCOM and to the Air Force's AFWERX. Meanwhile, as full disclosure looms, these specific attacks have resulted in "Pandora's box opening and curiosity of the use detrimental results of beamed weapon systems and devices attacks. An example of how portable systems work, is exampled by the off-the-shelf devices are designed to repulse vermin such as rats and insects from homes and have a range of about 30 to 40 feet.

The bottom line is that we are facing a threat, nationwide and globally as the use of these weapon explodes both officially and unofficially.

Balaban further stated, *It's the perfect kind of gray warfare. It's a very good way to degrade your opponent's capabilities so you can exploit it with conventional means,*

The fact is, targets in the US who report being used for human guinea pig experimentation, report that various types of beamed assaults arrive and many agree with the game plan and Standard Operating Procedure.

United States reported, Targeting also combined patented technologies that beamed noises and verbal communication using systems designed to focused on the cochlea of the ear, Many types of mind invasive, psychophysical, psychotronic technologies of which the military is testing, specifically the USAF and Navy, in my case at the helm of neighborhood experimentation of weapon systems, including, as stated previously, military microwave Active Denial System - Pain Ray beamed assaults, incorporating drones, antenna strategically, setup in compliant neighboring locations and vehicles..

Ultimately the team were in complete agreement that such weapons could be used clandestinely against a political leader, for example, to ultimately destabilize a society, as well as in ways that are disruptive on a variety of scales from psychophysical systems in the individual to systems in the social and political realm targeting, individuals, communities and large populations.

The US history of experimentation of human experimentation is an undeniable and documented fact that include the exposure of humans to many chemical and biological weapons (including infections with deadly or debilitating diseases) human radiation experiments, injections of various toxic and radioactive chemicals, surgical and medical experiments, with assistant from doctors, interrogation and torture experiments.

It includes various human testing that involved mind-altering substances, again, illegal drug infiltration into specific communities, along with unethical dispensing of powerfully addicting controlled substances easily accessible legally, and a wide variety of other experiments which are monitored and essentially are chemical warfare experimentation. There is also an official patents used on individuals, that can be used that is designed to keep people addicted. This is done by capturing the brainwave frequency during desire then beaming it back to bioelectric which had no firewall and body of targets. Many targets who have been used for experimentation since early childhood have all been taken through both illegal and legal drug experimentation that is monitored in real time for studies of their effect. The fact is Hitler used Meth to motivate his troops and was also addicted as well.

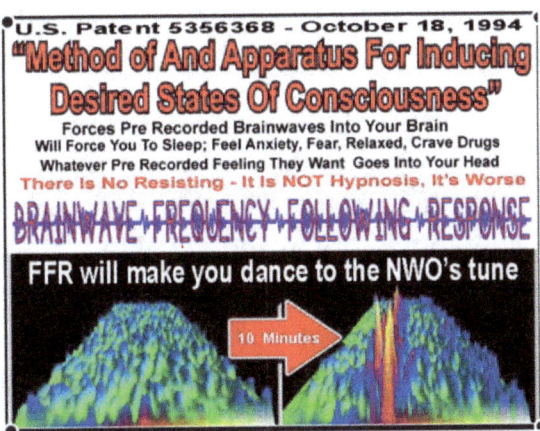

As stated previously many of these tests of various types are performed on children, the sick, and mentally disabled individuals, often under the guise of medical treatment. In many of these studies, a large portion of the subjects as it is today were poor, racial minorities, or prisoners or those caught up in the foster care system without families or family support and many drugged.

Many of these experiments violated US law. Some others were sponsored by government agencies or rogue elements, including the Centers for Disease Control, the United States military, and the Central Intelligence Agency, or they were sponsored by private corporations, DOD Contractors which were involved in military activities. The human research programs were usually highly secretive and as reported previously performed without the knowledge or authorization of Congress, and in many cases information about them was not released until many years after the studies had been performed. Based on many coming forth today, two things are clear, MKULTRA never ended and history is repeating itself, over and over again and people are coming forth before the experimentation is over is in this case it ever will be.

This program sends target's men, women and children through a living hell on Earth using various tactics, mind invasive technology, organized stalking, discrediting, destruction of families, isolation, subjugation torture and in some case high tech murder, if they can get away with it and not have the finger pointed at the real culprit. The fact is good and evil has been a part of the human experience since the beginning of time.

The cold-hard truth is that ongoing research, TESTING, and development programs continue. This has resulted in patented advanced technologies, patented at the United Patent and Trademark Office and mind invasive, behavioral modification, psychophysical beamed systems being tested across the USA today and globally.

When many say today the official's pulling the strings in every state in America are Fusion Center connected, connecting and have deployed their official servants looking for strategic methods of silencing, this is real.

It could mean using someone I reported, possibly, participating in million dollars' fraud around a loved one, and opening an official investigation who is then beamed anger to come after you to destroy you and them simultaneously this program's kill two birds with one stone technique. This was the techniques around Myron May's demise.

The fact is, the Assistant DA, Myron May was monitored up till his death bed as everyone is in real-time. Every move he took was monitored and coordinated which is typical of this massive space-based program.

Although he shot at three people, he missed two and injured one who recovered

To this day, many aware of this program and how people are used and influenced, believe that this and his focus directed specifically, on me alone, was actually a high-tech influenced setup to connect me with, the action he took, and then pretend I was involved or motivated it, which as I reported is actually what happened.

The best kept, heinous, official cover-up, is that our military are testing these weapon systems, which are portable, handheld, land, sea, drone and space-based, through the wall radar, on the civilian population and it reported by, again, highly credible individuals, nationwide.

Many have been added to a so-called *Watch-List* for justification and today basically for every reason under the Sun including Human Rights Advocacy and whistleblowing.

All then become human guinea pigs for official training, along with the expansions of police, and those involved will outright lie within the coverup on those they perceive a threat.

With that said, Targets are challenged to awaken

While spiritual awakening can be painful and difficult, especially under these intense circumstances, for many people, it is also a transformative and an enriching experience allowing you to cope with anything. True also, feeling lost, confused, alone, anxious during a spiritual awakening is entirely normal. There are certain stages within the spiritual awakening journey, each with its own set of experiences, challenges, and gifts. This program challenges your very Soul and endurance. Understanding these stages can help you better navigate your own spiritual path and provide invaluable life lessons along the way. Targets if you are to survive you must push for this, and begin the healing process.

It starts with the hardest thing of all, forgiveness after being treated so badly and suffering. No need to wish anything harmful on these evil people who chose this path as well. They are creating their own destiny and do not need your help by lowering your frequency.

Never doubt that the truth will prevail, slowly but surely.

For example, as shown below, recognition is beginning to unfold regarding brain targeting technology with Chile as the example.

Chile: Pioneering the protection of neurorights

Chile is set to become the first country in the world to legislate on neurotechnologies and include "brain rights" in its constitution.

21 March 2022 - Last update:13 October 2023

[Excerpt]

Lorena Guzmán H.

Science journalist, based in Santiago, Chile.

In 2021, Chile's Senate unanimously approved a bill to amend the constitution to protect brain rights or neuro-rights. The Chamber of Deputies reviewed and approved the amendment in September that year. It is now expected that the bill will be signed into law by the country's president.

Once the process is completed, Chile will become the world's first country to have legislation to protect mental privacy, free will and non-discrimination in citizens' access to neurotechnology. The aim is to give personal brain data the same status as an organ, so that it cannot be bought or sold, trafficked or manipulated.

At the same time, a constitutional reform to amend Article 19 of the Magna Carta, the country's constitution, is being considered to *protect the integrity and mental indemnity of the brain from the advances and capacities developed by neurotechnologies.*

The adoption of such a legal arsenal may seem premature in view of the development of neurotechnologies, which are still limited in their capacity to act on the human brain. But experts are already sounding the alarm and insisting on the need to legislate before intrusive applications become widespread – especially as progress in the field of neurotechnology continues to accelerate.

In April 2021, Neuralink, entrepreneur Elon Musk's brain-machine interface company, released a video of a monkey playing a video game after getting implants. The brain-machine interface technology used to do this is still in its infancy, but it opens the way to an infinite number of applications.

Chile could become the first country to pass legislation to protect mental privacy...

***

Without a doubt there is a global push for Artificial Intelligence and brain chipping

In an interview with Klaus Schwab's in 2016 with a Swiss French TV moderator, Schwab asked, *Imagine by 2025 we may all have a chip implanted somewhere in our body or brain, and we may be able to communicate with each other without a telephone, even without using our voice...?* He called this ability a fusion between the physical, digital, and biological world.

He spoke of also talks about servants' butlers" as robots, that are not just slaves, but rather assistants, as they function with Artificial Intelligence (AI), and will learn from us....

Noted in this interview is Schwab's obsession with the Fourth Industrial Revolution. It is defined as the full digitization of everything, and it seems to be boundless. See this full 2016 interview (video 28 min.), with the chipped humans beginning at 00:02:30. In the following link entitled WEF Obsession with A.I. and Brain Chipping

One thing is certain, we are undoubtedly moving towards globalization and a planned One World Government, and hope to drastically reduced world population to manageable numbers of humans on Earth. This apparently is the number one objective of those who have deemed themselves the Controller of humanity., It is termed "The Great Reset" and a reported United Nation's Agenda for 2030.

The global agenda of The Fourth Industrial Revolution, AI, and digitization of everything are just instruments to get there faster with global space-based psychophysical technology, and mass human experimentation setting the global stage for and mass social population control ushering the way.

Some believe that experimentation includes bioweapon tools such as COVID "vaccines", and new virus "X" – not yet existing, but roaming somewhere out ready to strike strategically for this globalist genocide as well as the tremendous climate hoax. Climate change has reportedly been the guise and lie to brainwash humanity that drastic measures must be taken for human survival which is used as a blueprint to justify population reduction.

In fact, the Club of Rome's devastating Report of "Limits to Growth" reportedly is still the blueprint for much of what is going on today likening climate change to a eugenics' dream which may be strategically realized. If we, the People, let them.

This is why every aspect of what is happening today demands awareness for it is easier to dupe the death, and blind.

Reference link:

https://www.globalresearch.ca/wef-obsession-ai-brain-chipping/5847563

> Never worry about who will be offended if you speak the truth. Worry about who will be misled, deceived and destroyed if you don't

Military Sources

Bioeffects of Selected Non-Lethal Weapons

Neuroelectric Activity and Analysis in Support of Direct Brainwave to Computer Interface Development

Radiofrequency Radiation Dosimetry Handbook

***

MKULTRA Subproject 119

Literature Review Papers *by American Biologist* and Christians Against Mental Slavery member

John J. Mc Murtrey (R.I.P)

***

NEW Microwave Vibration and Pulse Effects *NEW*

Technological Simulation of Hallucination

Project Bizarre Weapons Implications: Are Psychiatric Diagnosis, and Microwave Exposure Standards Presumptive?

A Simulated Hallucination Mechanism Compared to Hallucination Brain Response Studies

Inner Voice, Target Tracking, and Behavioral Influence Technologies

Thought Reading Capacity

Physiologic Word Recognition

Remote EEG Discussion

Recording Microwave Hearing Effects

Microwave Bioeffect Congruence With Schizophrenia

Remote Behavioral Influence Technology Evidence

Book Websites

bigbrotherwatchingus.com

Associated Blog

youarenotmybigbrother.blog

Beyond a shadow of a doubt humanity is dealing with a fast growing, masterful evil, literally in high places and it is no Joke!

What the Bible says about Literal Demonic Principalities
(From *Forerunner Commentary*)

The Bible uses *world* (*cosmos*) as man's system—of government, economics, religion, education, culture, etc.—established apart from the Creator God. This system is the source of much of what we believe and, along with its author, Satan, has been our god, though we did not realize it. Because Satan has been clever enough to include some of the true God's system, beliefs, stories, and practices within his, the Devil's system has an air of righteous authority. We can feel good, even joyous and inspired, while doing evil—like committing idolatry—in submitting ourselves as servants to his way.

We have been the unwitting slaves of an invisible, perversely intelligent, deceitful, powerful, and heartless master who is the God of this world (II Corinthians 4:4). He has created cultures with ways of life appealing to our self-centered natures. He stimulates our spirit through corrupt music, literature, art, and religion. He diverts our attention from more important concerns of life by means of entertainment with erotic visual and auditory impact. He has enslaved our minds by appealing to the desires of the flesh and of the eyes and the pride of life (I John 2:16) almost from the time we were born. He confuses us by hiding or shading the truth, denying absolutes, distorting reality, emphasizing vanity, and making available such a spectrum of opinions that disagreement is the standard operational feature of life. He pits us in competition against each other and makes us feel defensive, insecure, and untrusting.

By the time we are adults and God calls us, it takes a miracle mightier than God ever expended liberating the Israelites to even begin to free us from the demonic clutches of the pharaoh of this world, Satan the Devil!

John W. Ritenbaugh – *After Pentecost, Then What?*

Believe it or not!

References:

The Visual Pathways

Our eyes are 'wired' so that the left visual field of our total vision (that is, both eyes), goes to the right side of our brain and vice versa.

The neural pathways start with the rods and cones located at the back of the eye and end at the visual cortex at the back of our brain. Visual signals travel via synapses in the ganglion cells located behind the retina and leave the eye via the optic nerve. During this time the brain has already done a certain amount of visual analysis which delays the signal a few milliseconds.

The optic nerve from each eye splits into what is known as the optic chiasm which is the nerve network that routes the visual image from the right visual fields of both eyes (left half of each retina) to a nest of neurons called the left lateral geniculate and on to the visual cortex. The geniculate cells are connected to the thalamus by synapses forming the brain's sensory coordinating area or 'gateway'. Photic stimulation evokes potentials into the thalamus which then relays the information to the neo-cortex, as well as neighboring cortical areas and several targets deep within the brain.

The Auditory Pathways

The outer ears gather and transmits sound waves down the auditory canal where they reach the tympanic membrane and this vibrating membrane of the ear drum passes the vibration of the sound into the ossicles which consist of the malleus, incus and stapes (hammer, anvil and stirrup respectively).

The stapes fits into the oval window of the cochlea which responds to different pitches according to the location between the head and tail of the cochlea.

The cochlea contains the sensory elements called cilia which send the impulses into dendrites of nerve fiber neurons whose axons make up the fibers of the auditory portion of the VIII cranial (vestibular-cochlear) nerve.

These neurons are the first order neurons of the neural pathway, and proceed toward the brain stem where they form synapses with the cochlea nuclei located in the lower pons and upper medulla and consist of some 1 dozen different masses of cell bodies concentrated into three main groups of the olivary body.

The olivary body sends the auditory signals to the motor system of the ear and may be involved in reflexes. The inferior colliculus is involved in the creation of motor responses to auditory stimuli and the medial geniculate body which is located in the thalamus, serves as a relay station on the way to the auditory cortex. The auditory cortex lies in the temporal lobe, where low frequency sound is discriminated anteriorly, whilst high frequency sound is discriminated posteriorly. The auditory association areas surround the auditory cortex, and it is here that the brain integrates, remembers and analyses various types of sound input - it takes about 10 milliseconds for sound impulses to travel from the outer wear to the auditory cortex.

Sound induces an audio-evoked-response in the EEG and is a powerful psychological test of hearing because it indicates that stimulation by sound has caused a response in the auditory system and the brain. When the brain is exposed to rhythmic, evenly spaced on/off tones, brain wave following or 'entrainment' follows. Following usually occurs within seconds of exposure to the sound, and the 'trance' usually follows after about six minutes

(For more information on the use of sound as therapy, see the article on SAMONAS sound therapy)

Utilizing The Senses of Vision and Hearing to Affect The EEG

The senses of sight and hearing, by their very nature, provide a favorable means for affecting brainwaves. By presenting pulsed and sequenced audio and visual stimulation to the brain for a short period of time, the brain will begin to resonate or 'entrain' or 'follow' at the same frequency as the stimulus. This effect is commonly referred to as brainwave entrainment (BWE), or habituation, whereby the body and mind adapt to the stimulus.

In addition to entrainment, the imagery created by the visual and auditory stimulation provides a focus for the mind and helps to quieten internal dialogue or 'mind chatter'.

It is possible with AVE technology to recreate relaxation and meditation like that achieved by the *masters* and to experience that same peace and tranquility with only a half hour session using the techniques of audio visual and/or electromagnetic stimulus entrainment.

With the development of sophisticated electronic physiological measuring devices, scientists now conduct and record research to show the effects photic stimulation has on humans. Since the discovery of photic driving in 1934, many research articles have been printed in scientific and medical journals on the effects of BWE. In efforts to better understand the brain, most early research only observed the physiological effects of BWE directly and not the clinical benefits of BWE. It has only been more recently that clinical research has been conducted.[2.]

Neural Circuitry

Effective communication in the brain relies on neural circuitry. Neurons (nerve cells) consist of a cell body, axon and dendrites (the filament-like extensions of an axon).

In a normally functioning healthy brain, the dendrites of a given axon connect with the dendrites of many other axons, therefore, fostering full communication of information.

Often, development of neural pathways may be delayed, resulting in a pronounced deficits of many of the functions of hearing (CAPD) comprehension and/or sensory motor development. These delays may be global (pervasive developmental disorders) or can result in specific learning difficulties.

As ageing occurs, the brain loses some of its neural circuitry due to shrinkage and a reduction in the number of dendrites. The ability of neurons to communicate thus decreases as these connections are reabsorbed by the cell body with age or non-use.

Common symptoms (which can start as early as 40) include forgetting the names of people, and then names of things and facts, short-term memory deficits, difficulty following instructions and memorizing material.

However, we can slow down or even reverse this process, by providing multi-sensorial stimulation to our brains.

Stimulation can take the form of various types of neuro-feedback, bio-feedback, psychological counselling, healthy eating, exercise, Brain Wave entrainment (BWE) and a plethora of non-drug approaches that focus on prevention rather than correction.

When traumatic brain injury (TBI) occurs, connections between neural circuitry are also disrupted which leads to aberrant brain wave activity. This can occur through diffuse axonal shearing, or more directly as damage to specific brain areas.

Whilst the site of injury is unlikely to repair itself, (without stimulus the neuron is likely to undergo involution and eventually die), other neural pathways can be stimulated to be evoked by the brain, compensating for and perhaps in time performing the functions of those damaged. This inherent plasticity of the brain is what makes AVE and biofeedback so effective for many people.

With the use of an AVE unit a person can activate a pre-programmed AVE session at the simple press of a button to entrain the brain at a frequency that would be most useful for their particular needs or concerns.

Applications of AVE

Some of the uses and benefits of BWE with an AVE unit, as indicated by the research literature, are as follows:

- Exercises the Brain
- Enriched environment
- Increase in dendritic length
- Increase in brain mass
- Increase in blood supply
- Nerve cells are designed to receive stimulation resulting in growth and change. This growth and change are fundamental throughout life.
- Reduce Stress and Produce Deep Relaxation
- Decrease in stress related neurochemicals
- Reduce muscle tension
- Lower blood pressure
- Reduce heart rate
- Boost IQ
- LD average increase over 20 points and as high as 30
- ADD/ADHD - 12-20 increase
- Autism
- Increase alertness
- Reduce tantrums and aggressiveness
- Reduce hyperactivity
- Improved speech articulation and vocabulary
- Normalization of sleep patterns

- Accelerated Learning
- Learn more - learn faster - more adept at learning difficult and complex material due to increased dendritic outgrowth - more neural connections make learning easier
- Increase Memory
- Improve both, long and short term, memory
- Produce Peak Performance
- Produce high efficiency, effortless "flow" states
- Substance Abuse Problems
- Produce unprecedented recovery rates
- Overcome Depression and Anxiety
- Greatly reduce or eliminate chronic depression and anxiety
- Alleviate Pain
- Eliminate or greatly reduce both chronic and transient pain
- Boost Immune Function
- Increase the power of the immune system to overcome existing diseases and boost its resistance to infection

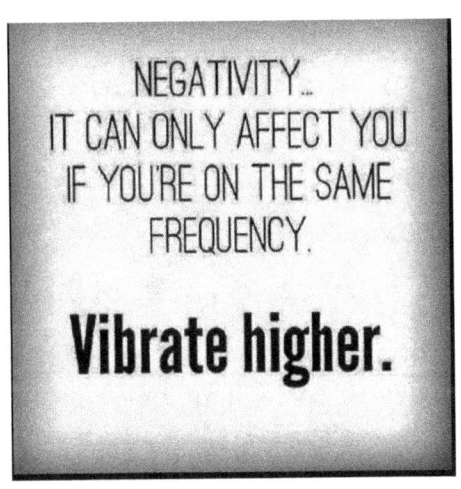

References:

Other papers and studies of AVE include:

Dental Research

The Effect of Repetitive Audio/Visual Stimulation on Skeletomotor and Vasomotor Activity in the Low Hypnotizable TMJ Subject, Dr. Norman Thomas B.D.S., B.Sc., Ph.D. & David Siever, C.E.T. Hypnosis, 1988.

A Technique for Rapidly Inducing Hypnosis, Bernard S. Margolis, D.D.S., CAL, June, 1966, Pages 21 to 24.

Tension Occurring in Muscles of Mastication During Jaw Opening - Research using Brainwave Entrainment Devices, David Siever, C.E.T., Unpublished.

The Application of Audio Stimulation and Electromyographic Biofeedback to Bruxism and Myofascial Pain-Dysfunction Syndrome, Ardeer Mains, Rudolf Miracles, and Hugo Adrian, Santiago, Chile,

Department of Physiology and Biophysics, Faculty of Medicine, University of Chile (1981).

Headache Research

The Treatment of Migraine with Variable Frequency Photo-Stimulation, D.J. Anderson, B.Sc., M.B., B.S. (1989).

Slow Wave Photic Stimulation in the Treatment of Headache - A Preliminary Report, Glen D. Solomon, M.D.

PMS, EEG Biofeedback, and Photic Stimulation, Presented at the 1995 SSNR Annual Conference, David Noton, Ph.D.

Pain Research

Living with Chronic Pain - A Holistic Treatment Program for Wellness, Frederic J. Boersma, Ph.D., University of Alberta (1990)

The Use of Repetitive Audio/Visual Entrainment in the Management of Chronic Pain, Frederic Boersma, Ph.D. and Constance Gagnon. M.Ed. (1992) Medical Hypnoanalysis Journal.

Relaxation Research

Mindworld Study #2 - Pilot Study - Effects of S/E Mediated Stress Management with the Metro Dade Police Department, Stress Reduction Study, Dr. Juan Abascal and Laurel Brucato (1989-1991) Miami Dade Community College. The Effects of Light and Sound Stimulation when used for Relaxation - Juan R. Abascal, Ph.D., and Larel L. Brucato, Ph.D. Miami - Dade Community College

Stress Reduction for Audio/Visual Integrated Stimulation and Self Therapy - Comptronic Devices Limited

Hypnosis in Anesthesiology, M.S. Sadove, M.D., Chicago, Illinois Medical Journal, July 1963, Pages 39 to 42.

Flicker Potentials and the Alpha Rhythm in Man, James Toman, Journal of Neurophysiology, 1941, Vol. 4, Pages 51 to 61.

Colorr Illusions and Aberrations During Stimulation by Flickering Light, W. Grey Walter, Nature, Vol. 177, Page 710.

Responses to Clicks from the Human Brain: Some Depth Electrographic Observations, Gian Emilio Chatrian, M.D., Magnus C. Petersen, M.D., and Jorge A. Lazarte, M.D. - Rochester State Hospital (1959).

Visual Evoked Responses Elicited by Rapid Stimulation, Jo Ann S. Kinney, Christine L. McKay, A.J. Mensch, and S.M. Luria, Naval Submarine Medical Research Laboratory, Naval Submarine Medical Centre, Naval Submarine Base New London, Groton, Connecticut (1972).

The Prognosis of Photosensitivity, P.M. Jeavons, A. Bishop, and G.F.A. Harding, Clinical Neurophysiology Unit, Department of Vision Sciences, Aston University, Birmingham, England (1986).

A Comparison of Depths of Relaxation Produced by Various Techniques and Neurotransmitters Produced by Brainwave Entrainment, Shealy and Forest Institute of Professional Psychology, Comprehensive Health Care, C. Norman Shealy, M.D., Ph.D., Roger K. Cady, M.D.,

Richard H. Cox, M.D., Ph.D., Saul Liss, William Clossen, Ph.D., Diane Culver Veehoff, R.N., Ph.D.

Influence of Color on the Photo Convulsive Response, T. Takahasi and Y. Tsukahara, Department of Neuropsychiatry and Department of Physiology, Tohoku University School of Medicine, Sendai, Japan (1976).

EEG Alpha Training, Hypnotic Susceptibility and Baseline Techniques, Bruce Crosson, Rodger Meinz, Eric Laur, Don Williams, and Ted Andreychuk, Texas Technological University (1977)

Altered States of Consciousness and Hypnosis: A Discussion, Erika Fromm, The University of Chicago (1977).

Hazard of Video Games in Patients with Light-Sensitive Epilepsy, Neil R. Dahiquist, MD; James F. Mellinger, MD; Donald W. Klass, MD (1983).

Megabrain Report - Recent Studies in Sound and Light, Julian Isaacs, Ph.D.

The Clinical Guide to Light/Sound Instrumentation & Therapy, Thomas H. Budzynski, Ph.D.

White and Red Lights in Photic Stimulation, David Siever, C.E.T.

Isochronic Tones and Brainwave Entrainment, David Siever, C.E.T.

An Electronic Aid for Hypnotic Induction: A Preliminary Report, William S. Kroger, M.D., and Sidney A. Schnieder, P.E.

Muscle and Fitness - Mind Over Matter = Muscle (article), Michael Hutchison (1993).

Fast Entry To Meditative States With Light and Sound Units, CMC, England

Auditory Beats in the Brain, Gerald Oster, Scientific

References

"Commentaries, Emerging Technology: Creator of Worlds," W. Michael Guillot, at:

"Brain-Computer Interfaces, U.S. Military Applications and Implications, An Initial Assessment" by Anika Benedick, Timothy Marler, Elizabeth M. Bartels,

"Emerging Cognitive Neuroscience and Related Technologies. National Research Council (US) Committee on Military and Intelligence Methodology for Emergent Neurophysiological and Cognitive/Neural Research in the Next Two Decades. Washington (DC): National Academies Press (US); 2008., at:

https://www.ncbi.nlm.nih.gov/books/NBK207933/

United States Army Research Institute for the Behavioral and Social Sciences, at:

https://ssl.armywarcollege.edu/dclm/pubs/Developing%20Army%20Strategic%20Thinkers.pdf

Laing, R.D. (1985) : *Wisdom, Madness and Folly: The Making of a Psychiatrist*. Macmillan, 1985

Welsh, Cheryl (1997): Timeline of Important Dates in the History of Electromagnetic Technology and Mind Control

Welsh, Cheryl (2001):Electromagnetic Weapons: As powerful as the Atomic Bomb, President Citizens Against Human Rights Abuse, CAHRA Home Page: U.S. Human Rights Abuse Report:

Begich, Dr. N. and Manning, J.: 1995 *Angels Don't Play this HAARP, Advances in Tesla Technology*, Earthpulse Press.

ZDF TV: "Secret Russia: Moscow – The Zombies of the Red Czars", Script to be published in *Resonance*, No. 35

Aftergood, Steven and Rosenberg, Barbara: "The Soft Kill Fallacy", in *The Bulletin of the Atomic Scientists*, Sept/Oct 1994.

Becker, Dr. Robert: 1985,*The Body Electric: Electromagnetism and the Foundation of Life*, William Morrow, N.Y.

Aback, Moir: International Movement for the Ban of Manipulation of The Human Nervous System and go to: Ban of Manipulation of Human Nervous System

"Is it Feasible to Manipulate the Human Brain at a Distance?"

"Psycho-electronic Threat to Democracy"

Nature: "Advances in Neuroscience May Threaten Human Rights", Vol, 391, Jan. 22, 1998, p. 316; (ref Jean- Pierre Changeux)

Space Preservation Act: Bill H.R.2977 and HR 3616 IH in 107th Congress – 2nd Session: Section 7

Delgado, Jose M.R: 1969. "Physical Control of the Mind: Towards a Psycho-civilized Society", Vol. 41, *World Perspectives*, Harper Row, N.Y.

US News & World Report: Lockheed Martin Aeronautics/ Dr. John Norseen; Report January 3/10 2000, P.67

Freud, Sigmund: 1919: *Art and Literature:" The Uncanny"*. Penguin, Also *"Those Wrecked by Success."*

Marks, John: 1988 :*The CIA and Mind Control – the Search for the Manchurian Candidate*, ISBN 0-440-20137-3

Persinger, M.A. "On the Possibility of Directly Accessing Every Human Brain by Electromagnetic Induction of Fundamental Algorythms"; *In Perception and Motor Skills*, June, 1995, vol. 80, p. 791 – 799

Tyler, J. "Electromagnetic Spectrum in Low Intensity Conflict," in "Low Intensity Conflict and Modern Technology", ed. Lt. Col. J. Dean, USAF, Air University Press, Centre For Aerospace Doctrine, Research and Education, Maxwell Air Force base, Alabama, June, 1986.

Rees, Martin *Our Final Century*: 2003, Heinemann. Conrad, Joseph: *The Secret Sharer*, 1910. Signet Classic. Maupassant, Guy de: *Le Horla*,

About the Author

Researcher, Author, Human Rights Advocate, Blogger, and Video Creator of a Docudrama related to Covert Human Experimentation and patented DOD *Voice of God* Synthetic Telepathy (RF Hearing Voices) discrediting human guinea pigs as Schizoid for decades! This woman, motivated from her heart, is an inspiration to many for her strength and endurance while navigating through a Goliath.

***

Declaration

"You must have control of the authorship of your own destiny. The pen that writes your life story must be held in your own hand." ~ Irene C. Kassorla

"Never be bullied into silence. Never allow yourself to be a victim. Accept no one's definition of your life; define yourself." – Harvey Fierstein

"The storm comes to make you grow!"

~ Renee Pittman

www.ingramcontent.com/pod-product-compliance
Lightning Source LLC
Chambersburg PA
CBHW072145070526
44585CB00015B/1003